W9-CSF-182

The Physics of
Moire Metrology

The Physics of
Moire Metrology

DR. ODED KAFRI
DR. ILANA GLATT

WILEY

A Wiley-Interscience Publication

John Wiley & Sons

New York / Chichester / Brisbane / Toronto / Sin

PHYSICS

04227219

Wiley Series in Pure and Applied Optics

The Wiley Series in Pure and Applied Optics publishes outstanding books in the field of optics. The nature of these books may be basic ("pure" optics) or practical ("applied" optics). The books are directed towards one or more of the following audiences: researchers in university, government, or industrial laboratories; practitioners of optics in industry; or graduate-level courses in universities. The emphasis is on the quality of the book and its importance to the discipline of optics.

Copyright ©1990 by John Wiley & Sons, Inc.

All rights reserved. Published simultaneously in Canada.

Reproduction or translation of any part of this work
beyond that permitted by Section 107 or 108 of the
1976 United States Copyright Act without the permission
of the copyright owner is unlawful. Requests for
permission or further information should be addressed to
the Permissions Department, John Wiley & Sons, Inc.

Library of Congress Cataloging in Publication Data:

Kafri, Oded.
   The phsyics of moire metrology/Oded Kafri. Ilana Glatt.
      p.   cm. — (Wiley series in pure and applied optics, ISSN
   0277-2493)
      Bibliography: p.
      1. Moiré method.   2. Optical measurements.   I. Glatt, Ilana.
   II. Title.   III. Series.

QC415.K34   1989   90-32480
535.4—dc20   CIP
ISBN 0-471-50967-1

Printed in the United States of America

10 9 8 7 6 5 4 3 2 1

Contents

QC 415
K34
1990
PHYS

v

Preface—Historical Background

The moire effect denotes a fringe pattern formed by the superposition of two grid structures of similar period. It attracts the observer's eye in everyday objects such as the folds of a lace curtain, the railings of a bridge, two overlapping nets, or the noise in a TV picture when the pattern photographed has a period similar to that of the monitor or camera. The attraction to this effect results from a fringe movement that is much faster than the object motion and the periodical shape of the effect. The beauty of moire fringes is the reason for their use in graphics, art, and clothing. The name probably originated from the textile world where it was first observed in the mohair fabric. Another suggestion is that moire is the French word for the *watered or wavy appearance* that is observed when layers of silk are pressed together at an angle so that a moire pattern is formed.

The first proposal to use the moire effect for scientific purposes was over 100 years ago when Lord Rayleigh [1] suggested it for grating analysis. Moire patterns resulting from circular and radial gratings were described by Righi [2] about 10 years later. Nevertheless, as far as we know, the moire effect was not utilized until the 1920s when Ronchi [3], Raman [4], and Datta [5] published their work. The next appearance of moire was in the 1940s when the application of in-plane strain analysis was suggested [6, 7]. In 1956, Guild [8] published a book on the theory of using the moire effect for grating analysis and from this period on, the subject was no longer neglected. In 1969 and 1970 both Theocaris [9] and Durelli and Parks [10] published books dedicated to strain analysis using the moire effect. As a result of the extensive study of strain, several mathematical tools and experimental techniques were developed, such as shearing moire [11–13] and the indicial equation formalism [14, 15].

Moreover, the scope of moire analysis was extended from strain analysis to the analysis of large diffusive objects [16–19]. It is important to mention the impact of the work done by Takasaki (which is similar to that of Meadows et al.), which created a new field of topographic contouring of the human body for medical purposes. An additional important application is the mapping of phase objects and specular surfaces [20–23].

There are many more applications, which we will not mention, in medicine, crystallography, and even optics. The purpose of this book is not to provide a historical summary on what has been done in this field, in which some 1000 scientific papers have been published, nor to discuss any specific technique. On the contrary, we would like to put together all the steps of the past 100 years, for the first time, as one coherent piece and to establish moire technology as a chapter in the optics section of physics. We adopt a point of view, which to our knowledge is new in this field, of treating the gratings used in moire analysis as an artificial analog to electromagnetic waves. Therefore, we can compare moire analysis with conventional optical methods based on wave properties (i.e., interferometry). We will show that for every interferometric technique in metrology, there is an analogous one in moire metrology and vice versa. We feel that the scientists involved in optical metrology have a real choice between interferometric and moire methods. Unfortunately many of them have only a vague idea about the moire alternative and consider it inaccurate, not quantitative, etc. No single textbook in optics, to our knowledge, provides a comprehensive discussion of moire analysis, albeit moire analysis is a real alternative to interferometry and holography.

This book is intended for physicists interested in optical metrology and optical engineers. It provides the basic tools that a physicist or optical engineer needs to enter into research in the field. As mentioned, this book is not a review, and therefore we would like to apologize to all the scientists whose contributions are not mentioned. It does not mean that we think their contributions are not important enough, but only that they do not fit our analogy between gratings and light waves. We would like to thank our leading colleagues who helped us with material and advice: Olof Bryngdahl, Shunsuk Yokozeki, Daniel Post, R. Ritter, A. J. Durelli, C. J. Parks, and G. L. Rogers. At the end of each chapter, we have added a reference list of papers which we recommend to the reader who is interested in a wider scope than provided.

Special thanks to Kathi Kreske and Eliezer Keren for careful reading of the manuscript and many useful suggestions.

ODED KAFRI
ILANA BRÖNNIMANN-GLATT

Beer Sheva, Israel
Buchs (ZH), Switzerland
October, 1989

References

1. Lord Rayleigh, *Philos. Mag.* **47**, 81 (1874); **47**, 193 (1874).

2. A. Righi, *Nuovo Cimento* **21**, 203 (1887); **22**, 10 (1888).

3. V. Ronchi, *Attualita Scintifiche*, No. 37, N. Zanidelli, Bologna, 1925, Chap. 9.

4. C. V. Raman and S. K. Datta, *Trans. Opt. Soc.* **27**, 51 (1925–1926).

5. S. K. Datta, *Trans. Opt. Soc.* **28**, 214 (1926–1927).

6. G. A. Brewer and R. B. Glassco, *J. Aero. Sci.* **9**, No. 1 (1941).

7. R. Weller and B. M. Shephard, *Proc. Soc. Experimental Stress Analysis* **6**, 35 (1948).

8. J. Guild, *The Interference System of Crossed Diffraction Gratings*, Clarendon, Oxford, 1956.

9. P. S. Theocaris, *Moire Fringes in Strain Analysis*, Pergamon, London, 1969.

10. A. J. Durelli and V. J. Parks, *Moire Analysis of Strain*, Prentice-Hall, Englewood Cliffs, N.J., 1970.

11. F. K. Ligtenberg, *Proc. SESA* **12**, 83 (1954).

12. J. P. Duncan and P. S. Sabin, *Exp. Mech.* **5**, 22 (1965).

13. S. Yokozeki and T. Suzuki, *Appl. Optics* **9**, 2804 (1970).

14. G. Oster, M. Wasserman, and C. Zwerling, *J. Opt. Soc. Am.* **54**, 169 (1964).

15. S. Yokozeki, *Opt. Commun.* **11**, 378 (1974).

16. P. S. Theocaris, *J. Sci. Instrum.* **41**, 133 (1964).

17. C. Chiang, *Brit. J. Appl. Physics* (*J. Phys. D*) Sec. 2 **2**, 287 (1969).

18. D. M. Meadows, W. O. Johnson, and J. B. Allen, *Appl. Optics* **9**, 942 (1970).

19. H. Takasaki, *Appl. Optics* **9**, 1467 (1970).

20. Y. Nishijima and G. Oster, *J. Opt. Soc. Am.* **54**, 1 (1964).

21. S. Yokozeki and T. Suzuki, *Appl. Optics* **10**, 1575 (1971).

22. A. W. Lohmann and D. E. Silva, *Optics Commun.* **2**, 1690 (1971).

23. O. Kafri, *Opt. Lett.* **5**, 555 (1980).

1

Introduction to Moire Metrology

1.1. OPTICAL METROLOGY AND LIGHT

Metrology is the science of weights and measures. Measuring dimensions, either of space or of time, is performed by comparing the size of an unknown interval to a standard reference. In measuring time the preferred references for calibration since early civilization have been the periodic phenomena of nature. For example, the harmonic oscillation of a pendulum is used to measure the duration of short events, whereas the spinning of the Earth about its axis and the Earth's motion in its planetary orbit measure time gaps of days and years.

Light, an electromagnetic periodic perturbation propagating in space at a constant velocity, provides an ideal measuring device of both time and distance. The periodic nature of light was probably first suggested in the seventeenth century by Robert Hooke, who described light as a rapid vibratory motion of the medium propagating at a very great speed. But the edifice to modern wave theory of light was no doubt laid by Christian Huygens [1] (1629–1695) and later revised in the nineteenth century by Thomas Young. Young suggested the transverse propagation of light waves, which was later established by James C. Maxwell as the electromagnetic field [2]. Young also correctly interpreted the interference effect caused by the coincidence of two light waves [3]. This fringe pattern is observed when two or more beams originating from the same source overlap, and it is the basis of the most important tool in optical metrology, the interferometer. One of the first interferometers was built by Michelson [4] in the 1870s. The Michelson interferometer, still widely in use, is shown in Fig. 1.1. The light emitted by an extended source is split into two beams by a half silvered mirror. The two beams

1

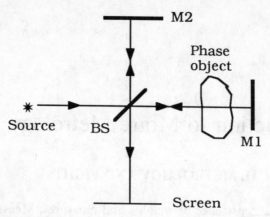

Figure 1.1. A Michelson interferometer. B.S. is the beamsplitter and M_1 and M_2 are the two mirrors.

propagate at right angles and are reflected back by two mirrors. The beams are mixed again by the same beam splitter and form an interference pattern on the mat screen. Since interferometry and interference effects are frequently referred to in this book, we now give a brief review on wave theory and, in particular, the superposition of waves [5].

1.2. LIGHT WAVES AND INTERFEROMETRY

Following the Maxwell formulation of electrodynamics, light is an electromagnetic wave, namely, its propagation is a solution to the differential wave equation [6]. Moreover, any linear combination of functions that are solutions of that equation also conforms a solution. Therefore, light can be regarded as a superposition of monochromatic plane waves, i.e., harmonic waves with planar wavefront, expressed by the complex function Ψ:

$$\Psi(\mathbf{r}, t) = A_0 e^{i(\mathbf{kr} \pm \omega t)}, \tag{1.1}$$

where $\mathbf{r} = [x, y, z]$ is the position vector relative to an arbitrary origin 0 and $\mathbf{k}$ is the wave vector perpendicular to the wave fronts.

The magnitude of $\mathbf{k}$ is the wave number $|k| = 2\pi/\lambda$, where λ is the wavelength of light. ω is the angular frequency of light given by $2\pi\nu$ ($\nu = v/\lambda$ where v is the speed of light in the medium), t is the time coordinate, and A_0 denotes the amplitude of the electromagnetic field.

When two harmonic waves are superimposed, the combined wave is the arithmetic sum of the two functions. Let us describe Ψ_1 and Ψ_2 as

$$\Psi_1 = A_{01}e^{i(\mathbf{k}_1\mathbf{r} \pm \omega t + \varepsilon_1)} \equiv A_1 e^{\pm i\omega t},$$
$$\Psi_2 = A_{02}e^{i(\mathbf{k}_2\mathbf{r} \pm \omega t + \varepsilon_2)} \equiv A_2 e^{\pm i\omega t}, \tag{1.2}$$

where ε_1 and ε_2 are constant phase terms and A_1 and A_2 are defined as the complex amplitudes of the waves. The amplitude of the composite wave formed by the superposition is the sum of the complex amplitudes, namely,

$$A = A_1 + A_2 = A_{01}e^{i\alpha_1} + A_{02}e^{i\alpha_2}, \tag{1.3}$$

where $\alpha_i = \mathbf{k}_i\mathbf{r} + \varepsilon_i$. The energy flux of a wave I is proportional to the square of the amplitude of the electric field, i.e., to AA^* (A^* is the complex conjugate of A). Thus

$$I \propto |A|^2 = AA^* = \left(A_{01}e^{i\alpha_1} + A_{02}e^{i\alpha_2}\right)\left(A_{01}^*e^{-i\alpha_1} + A_{02}^*e^{-i\alpha_2}\right). \tag{1.4}$$

Since A_{01} and A_{02} are real ($A_{0i} = A_{0i}^*$), the square complex amplitude is given by

$$|A|^2 = A_{01}^2 + A_{02}^2 + A_{01}A_{02}\left[e^{i(\alpha_1-\alpha_2)} + e^{-i(\alpha_1-\alpha_2)}\right]$$
$$= A_{01}^2 + A_{02}^2 + 2A_{01}A_{02}\cos(\alpha_1 - \alpha_2). \tag{1.5}$$

In words, the energy flux resulting from the superposition of monochromatic waves is the sum of the fluxes of the separate waves plus an interference term of the form

$$A_1A_2^* + A_1^*A_2 = 2A_{01}A_{02}\cos(\mathbf{k}_1\mathbf{r} - \mathbf{k}_2\mathbf{r} + \varepsilon_1 - \varepsilon_2)$$
$$\equiv 2A_{01}A_{02}\cos\delta, \tag{1.6}$$

where δ denotes the phase difference between the two interfering waves and $\cos \delta$ varies between $+1$ (for even multiples of π) and -1 (for odd multiples). The overall flux modulates between the two extreme positions called *total constructive* and *total destructive* interference for zero phase shift and phase shift of π, respectively. This gives rise to an interference pattern of alternating dark and bright fringes, whose spacing is proportional to the differences in optical paths swept by the two waves, in units of λ. The optical path difference (OPD) is introduced by the term **kr** ($|k| = 2\pi \nu n/c$, where c is the speed of light in vacuum and n is the refractive index of the medium).

In homogeneous media, the OPD responsible for the interference pattern is due to a difference in the geometrical path length. On passing through media of varying density the OPD reflects also the difference in n. Therefore, interferometric measurements can be applied to a wide variety of phenomena that affect the OPD. The applications include metrological as well as wavelength-dependent (spectroscopic) measurements (like Fourier transform spectroscopy or anomalous dispersion). The latter, however, are beyond the scope of this book.

1.3. THE MOIRE EFFECT

In this book we present a totally different approach to optical metrology than that presented by interferometry. Our approach employs the moire effect. Unlike interferometry, which is based on the wave nature of light and requires mutual temporal coherence, moire metrology uses light only for illumination. In principle, any source of light can be used. The term *moire effect* refers to a geometrical interference formed when two transmitting screens of similar motif partly overlap.

Throughout the text an analogy will be drawn between moire and interferometric techniques. We will show that in spite of the categorical difference between the two methods, quite similar or complementary data can be derived. Moreover, excluding the spectroscopic measurements, the innovative moire-based methods can replace interferometry in practically all metrological measurements, with a compatible degree of accuracy that is ultimately subject to the universal physical limitations.

1.4. ANALYSIS OF THE MOIRE PATTERN BY THE INDICIAL EQUATION FORMALISM

Although moire patterns can be observed whenever two transparencies of similar motifs are brought to partly overlap, we are specifically concerned with the linear transmission gratings shown in Fig. 1.2. When two such gratings of equal band spacing are superimposed, maintaining a slight intersection angle θ between the stripe directions, a straight fringe pattern is observed. The direction of the fringes is perpendicular to the bisector of the angle θ. The fringe spacing increases as the intersection angle is reduced. If the gratings are not identical, for example, one of the gratings is locally deformed relative to the other, the fringe pattern will reflect this deformity by an appropriate fringe shift.

The fringe profile is determined by the line shape of the gratings. In interferometry, for example, the fringe profile is proportional to a

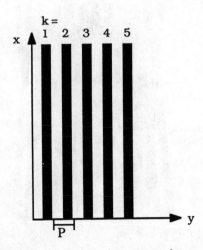

Figure 1.2. A Ronchi ruling consisting of opaque and transparent stripes of equal width and with pitch p.

square of $\cos \delta/2$ [assuming $A_{01} = A_{02}$ in Eq. (1.6)]. In moire techniques, which apply the square wave gratings of equally wide opaque and transparent stripes (known as Ronchi rulings), the fringe profile is triangular. The transmitted intensity is proportional to the phase shift between the two gratings (within 0 and $\pi/2$) and the tendency is reversed in the second quadrant (between $\pi/2$ and π).

Figure 1.3 is an enlarged section of a moire pattern obtained with two Ronchi rulings. As shown, a dark fringe is an imaginary line connecting all zones where light is blocked by interference of the opaque stripes, which are mutually translated by half a period (given by the pitch p). The bright fringes are the geometrical locus of the intersection point of the opaque stripes. The transmitted intensity averaged over a bright fringe is one-half of the transparent stripe transmittance and it decreases linearly toward total opacity along the dark fringe. When magnified it is seen that the moire effect is just an optical illusion, in which the bright fringes are chains of alternating transparent and opaque rhombs. But, when dense gratings of 6 lines/mm and above are used, the averaging done by the eye or by a detection system smooths the fringe pattern into continuous stripes.

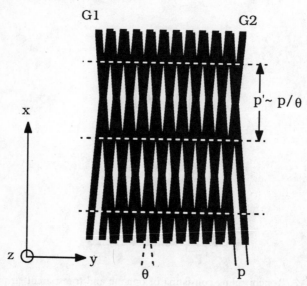

Figure 1.3. The moire pattern of two Ronchi rulings superimposed at a small angle θ

We use the mathematical formalism of indicial equations to indicate the fringe position. In this formulation, a grating is substituted by a discrete array of lines described by a set of equations, each specifying the location of a line in the array. For example, the equation for a Ronchi ruling of pitch p along the y axis is

$$y = lp, \qquad l = 0, \pm 1, \pm 2, \pm 3 \dots . \tag{1.7}$$

A moire pattern formed by the superposition of two line gratings, G_1 and G_2, is shown in Fig 1.4. The lines of grating G_1 form an angle $\theta/2$ with the x axis, and the grating G_2 is symmetrically inclined at $-\theta/2$. The gratings G_1 and G_2 are described by the two sets of equations, respectively,

$$y \cos \theta/2 = x \sin \theta/2 + np, \qquad n = 0, \pm 1, \pm 2, \dots ,$$
$$y \cos \theta/2 = -x \sin \theta/2 + mp, \qquad m = 0, \pm 1, \pm 2, \dots . \tag{1.8}$$

The geometrical loci of all points of intersection of the two gratings form an array of bright fringes whose index is given by

$$l = m - n \qquad l = 0, \pm 1, \pm 2 \dots . \tag{1.9}$$

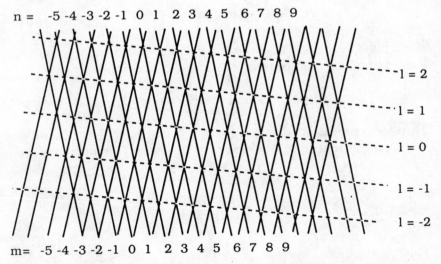

Figure 1.4. The indexing used in the indicial equations.

Substituting for m and n yields

$$l = \frac{2x \sin \theta/2}{p}. \tag{1.10}$$

By rearranging Eq. (1.10) we obtain the indicial equations describing the fringe pattern:

$$x = \frac{lp}{2 \sin \theta/2}. \tag{1.11}$$

For small values of θ ($\sin \theta \cong \theta$, in radians) we obtain

$$x \cong \frac{lp}{\theta}. \tag{1.12}$$

Equation (1.12) represents a set of straight lines perpendicular to the x axis, whose spacing is θ^{-1} times the pitch of the original gratings.

The formation of a moire pattern by superposition of linear grids resembles the interference pattern of two mutually inclined plane waves of equal frequency. Assuming that Ψ_1 and Ψ_2 are inclined by angles $\theta/2$ and $-\theta/2$ to the x axis, respectively,

$$\mathbf{k}_1 = \mathbf{j}k \cos \theta/2 + \mathbf{i}k \sin \theta/2,$$
$$\mathbf{k}_2 = \mathbf{j}k \cos \theta/2 - \mathbf{i}k \sin \theta/2, \tag{1.13}$$

where $\mathbf{j}$ and $\mathbf{i}$ are unit vectors in the y and x directions, respectively. The interference term contains the cosine expression

$$\cos(\mathbf{k}_1\mathbf{r} - \mathbf{k}_2\mathbf{r}) = \cos(2kx \sin \theta/2). \tag{1.14}$$

Therefore, similar to the moire effect, the spacing between bright interference fringes, i.e., $2kx \sin \theta/2 = 2\pi m$ (m is an integer), where $k = 2\pi/\lambda$, is

$$\frac{\lambda}{2 \sin \theta/2}.$$

The moire pattern, however, is not thoroughly analogous to the wave interference pattern. The oscillating part of the interference term consists only of the phase difference between the two waves, given by $k_1r - k_2r$. The moire pattern, on the other hand, contains both difference and sum terms. Since the difference pattern, which consists of the $l = m - n$ fringes, is easily discernible by the eye, we often ignore the coexisting high frequency moire fringes conforming to $l' = m + n$. Those fringes are parallel to the x axis and maintain

$$y = \frac{(m + n)p}{2\cos\theta/2} \cong \frac{l'p}{2}, \qquad (1.15)$$

namely, their frequency is about twice that of the Ronchi ruling. Nevertheless, when dense gratings ($p < 0.1$ mm) are used, these fringes are hardly observable by the naked eye or on a coarse grain photographic film. To completely remove these high frequency fringes so that we are certain they do not interfere with the measurement, an analogy is made with interferometry.

The light waves discussed so far are traveling waves, i.e., waves that move in space. But a transmission grating, as used in the moire effect, is closer in concept to a stationary wave. A stationary wave can be described as superposition of two harmonic waves of the same frequency propagating in opposite directions as

$$\Psi_1 = A_0 e^{i(\mathbf{k}\mathbf{r} + \omega t)},$$
$$\Psi_2 = A_0 e^{i(-\mathbf{k}\mathbf{r} + \omega t)}, \qquad (1.16)$$

the superposition of which results in

$$\Psi = A_0 e^{i\omega t} \cos \mathbf{k}\mathbf{r}. \qquad (1.17)$$

In a stationary wave, the amplitude ($A_0 \cos \mathbf{k}\mathbf{r}$) is constant in space and varies harmonically in time. The energy flux of the interference pattern formed by superposition of two stationary waves with amplitudes of $A_1 \cos \mathbf{k}_1\mathbf{r}$ and $A_2 \cos \mathbf{k}_2\mathbf{r}$ is given by

$$\begin{aligned} I &= A_1 A_1^* \cos^2 \mathbf{k}_1\mathbf{r} + A_2 A_2^* \cos^2 \mathbf{k}_2\mathbf{r} \\ &\quad + (A_1 A_2^* + A_2 A_1^*)\cos \mathbf{k}_1\mathbf{r} \cos \mathbf{k}_2\mathbf{r} \\ &= |A_1|^2 \cos^2 \mathbf{k}_1\mathbf{r} + |A_2|^2 \cos^2 \mathbf{k}_2\mathbf{r} \\ &\quad + \tfrac{1}{2}(A_1 A_2^* + A_2 A_1^*)[\cos(\mathbf{k}_1 + \mathbf{k}_2)\mathbf{r} + \cos(\mathbf{k}_1 - \mathbf{k}_2)\mathbf{r}], \qquad (1.18) \end{aligned}$$

i.e., in stationary waves the interference term consists of both difference and sum contributions. The conclusion that can be immediately
drawn from this analogy between stationary waves and transmittance gratings is that the sum contribution in the moire effect can be
eliminated by creating a traveling-wave grating. The high frequency
moire pattern is removed if the two gratings are moving in unison at
constant speed during the measurement, as was suggested by Allen
and Meadows [7].

1.5.   LINE SHAPE ANALYSIS OF MOIRE FRINGES

The line shape of the moire fringes formed by two identical gratings
is mathematically described by the autocorrelation function of the
grating's transmittance function. We describe the transmittance T of
an infinitely large square wave grating positioned at the y, x plane
with its stripes parallel to the x axis by

$$T(y,0) = \text{rect}\left(\frac{2y}{p}\right) * \frac{1}{p}\text{comb}\left(\frac{y}{p}\right). \qquad (1.19)$$

The rectangle function $\text{rect}(2y/p)$ acquires the value 1 for $|y| < p/4$
and 0 elsewhere. The comb function represents an array of equally
spaced δ functions, namely,

$$\text{comb}\left(\frac{y}{p}\right) = p \sum_{n=-\infty}^{\infty} \delta(y - np) = \sum_{n=-\infty}^{\infty} \delta\left(\frac{y}{p} - n\right), \quad (1.20)$$

where p is the spacing and δ is the Dirac δ function. The convolution of two functions, $f(x)$ and $g(x)$, denoted in symbols as
$f(x) * g(x)$, is defined by the integral

$$f(x) * g(x) = \int_{-\infty}^{\infty} f(x')g(x - x')\, dx'. \qquad (1.21)$$

Applying this formula, the transmittance functions of two Ronchi
rulings rotated by $\theta/2$ and $-\theta/2$ relative to the x axis are given by
T_1 and T_2, respectively:

$$T_1(y, x) = \frac{1}{p}\mathrm{comb}\left(\frac{r'}{p}\right) * \mathrm{rect}\left(\frac{2r'}{p}\right),$$

$$T_2(y, x) = \frac{1}{p}\mathrm{comb}\left(\frac{r''}{p}\right) * \mathrm{rect}\left(\frac{2r''}{p}\right), \qquad (1.22)$$

where

$$r' = y\cos\theta/2 - x\sin\theta/2 \quad \text{and} \quad r'' = y\cos\theta/2 + x\sin\theta/2.$$

In the indicial equation notation, the gratings G_1 and G_2 are described by

$$r' = np, \qquad n = 0, \pm 1, \pm 2, \ldots,$$
$$r'' = mp, \qquad m = 0, \pm 1, \pm 2, \ldots, \qquad (1.23)$$

respectively. Therefore, the bright fringes are parallel to the y axis and given by

$$x = \frac{(m - n)p}{2\sin\theta/2}. \qquad (1.24)$$

As one moves away from a bright fringe along the x axis, the overall transmittance decreases linearly with the distance. After half a fringe, a minimum is reached and the transmittance starts to rise again.

The overall transmittance of the moire pattern $T(y, x)$ is given by the product of the transmittance function of the two overlapping gratings:

$$T(y, x) = T_1(y, x)T_2(y, x) = \mathrm{rect}\left(\frac{2r'}{p}\right)\mathrm{rect}\left(\frac{2r''}{p}\right). \quad (1.25)$$

Since $r' - r'' = 2x\sin\theta/2$, one can rewrite Eq. (1.25) as

$$T(y, x) = \mathrm{rect}\left(\frac{2r''}{p}\right)\mathrm{rect}\left(\frac{2r'' - 4x\sin\theta/2}{p}\right). \qquad (1.26)$$

In practice, the observable transmittance T_{lp} of the moire pattern is a smooth function, i.e., a low pass filtered function of $T(y, x)$. This is achieved by averaging over one pitch (projected on the y axis) as

$$
\begin{aligned}
T_{lp}(y', x) &= \frac{\cos \theta/2}{p} \int_{y'-p/(2\cos\theta/2)}^{y'+p/(2\cos\theta/2)} T(y, x)\, dy \\
&= \frac{\cos \theta/2}{p} \int_{y'-p/(2\cos\theta/2)}^{y'+p/(2\cos\theta/2)} \mathrm{rect}\left(\frac{2y + 2x\tan\theta/2}{p/\cos\theta/2}\right) \\
&\quad \times \mathrm{rect}\left(\frac{2y - 2x\tan\theta/2}{p/\cos\theta/2}\right)\, dy,
\end{aligned}
\tag{1.27}
$$

namely, the overall transmittance function T_{lp} is an autocorrelation function of the grating's transmittance. This conclusion can be applied to any form of linear gratings, given a stripe transmission function $f(y, x)$:

$$
T_{lp}(y', x) = \frac{1}{b} \int_{y'-b/2}^{y'+b/2} f(y, x) f\left(y + 2x\tan\frac{\theta}{2}, x\right) dy, \tag{1.28}
$$

where b is the projection of the pitch on the y axis.

In the case of a Ronchi ruling, using the small angle approximation ($\cos\theta \sim 1$, $\sin\theta \sim \theta$), the solution is straightforward. We can rewrite Eq. (1.27) as

$$
\begin{aligned}
T_{lp}(y', x) &\cong \frac{1}{p} \int_{y'-p/2}^{y'+p/2} \mathrm{rect}\left(\frac{2y + x\theta}{p}\right) \mathrm{rect}\left(\frac{2y - x\theta}{p}\right) dy \\
&= \frac{1}{p} \int_{x\theta/2-p/4}^{-x\theta/2+p/4} dy
\end{aligned}
\tag{1.29}
$$

or

$$
T_{lp}(y', x) = \frac{1}{2} - \frac{x\theta}{p}, \tag{1.30}
$$

namely, the observed fringe profile is a triangular wave with already known periodicity of

$$p' \cong \frac{p}{\theta}.$$ (1.31)

The intensity varies between $\frac{1}{2}$ at the maximum where the lines intersect and zero at

$$z \cong \frac{p}{2\theta},$$ (1.32)

the center of the dark fringe.

1.6. MOIRE PATTERN RESPONSE TO GRATING'S DEFORMATIONS

After introducing the moire effect and the mathematical tools used for fringe analysis, it is an appropriate time to demonstrate how the moire effect is applied to optical metrology. In the preceding section we dealt with perfect gratings that form a flawless straight fringe pattern that is a magnified replica of the gratings. How then would a fringe pattern look if one of the gratings is slightly distorted? First, let us assume that one of the gratings G_1 is slightly shifted, parallel to the grating's direction in the direction of the increasing indicial number, by an amount δp $(\delta p < p)$ as shown in Fig. 1.5. The indicial equations for G_1 and G_2 are, respectively,

$$y \cos \theta/2 = x \sin \theta/2 + np + \delta p,$$
$$y \cos \theta/2 = -x \sin \theta/2 + mp.$$ (1.33)

The fringe position is now given by

$$x = \frac{(m-n)p}{2 \sin \theta/2} + \frac{\delta p}{2 \sin \theta/2} \cong \frac{lp}{\theta} - \frac{\delta p}{\theta},$$ (1.34)

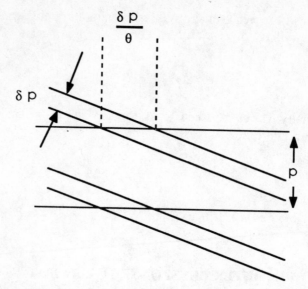

Figure 1.5.   Fringe shift resulting from translation of one grating.

namely, the fringes are shifted down by an amount θ^{-1} times the original grating's shift. A similar shift of the grating in the direction of the lines has no effect since $\delta p \ll p'$.

If one of the gratings G_1 is slightly rotated about the optical axis by an angle ϕ (clockwise, for this discussion), the indicial equations describing the situation are

$$y \cos(\theta/2 + \phi) = x \sin(\theta/2 + \phi) + np,$$
$$y \cos \theta/2 = -x \sin \theta/2 + mp. \tag{1.35}$$

The easiest way to solve this set of equations is by rotating the coordinate system clockwise by $\phi/2$ so that the gratings are inclined symmetrically to the new x' axis, namely,

$$y' \cos\left(\frac{\theta + \phi}{2}\right) = x' \sin\left(\frac{\theta + \phi}{2}\right) + np,$$
$$y' \cos\left(\frac{\theta + \phi}{2}\right) = -x' \sin\left(\frac{\theta + \phi}{2}\right) + mp. \tag{1.36}$$

The fringes are positioned perpendicular to x'

$$x' = \frac{lp}{2\sin((\theta + \phi)/2)} \cong \frac{lp}{\theta + \phi}. \qquad (1.37)$$

The fringes are rotated by $\phi/2$ relative to the initial case and are also denser than before the rotation by a factor $(\theta + \phi)/\theta$.

Now assume that the grating G_1 is slightly stretched by δp, as shown in Fig. 1.6. The resultant fringe distortion $\delta p'$ is given by

$$\delta p' = \frac{\delta p}{2\sin\theta/2}, \qquad (1.38)$$

as can be seen in the drawing.

Another important illustration of the analogy between electromagnetic wave interference and moire patterns is the *beats* phenomenon that occurs when two waves of slightly different frequencies interfere. This phenomenon can be seen from Eq. (1.5) by substituting

$$\alpha_1 - \alpha_2 = (\omega_2 - \omega_1)t, \qquad \omega_2 \neq \omega_1. \qquad (1.39)$$

The amplitude of the envelope is oscillating at the difference frequency, namely,

$$|A|^2 = A_{01}^2 + A_{02}^2 + 2A_{01}A_{02}\cos 2\pi ct\left(\frac{1}{\lambda_2} - \frac{1}{\lambda_1}\right). \qquad (1.40)$$

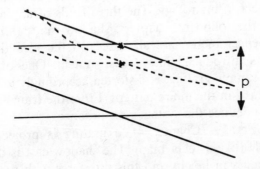

Figure 1.6. Fringe shift resulting from a distorted grating.

In the corresponding moire case we combine two parallel gratings

$$y = np_2,$$
$$y = mp_1, \qquad\qquad (1.41)$$

where $p_1 = kp_2$. Equation (43) can be written as

$$y = np_2,$$
$$y/k = mp_2. \qquad\qquad (1.42)$$

This yields,

$$l = y\left(\frac{1}{p_2} - \frac{1}{p_1}\right), \qquad l = n - m, \qquad (1.43)$$

which is the beats phenomenon in moire. In other words, a pattern in the direction of the original grating will have a pitch p given by

$$\frac{1}{p} = \frac{1}{p_2} - \frac{1}{p_1}. \qquad\qquad (1.44)$$

This is analogous to the beats wavelength in interferometry,

$$\frac{1}{\lambda} = \frac{1}{\lambda_2} - \frac{1}{\lambda_1}. \qquad\qquad (1.45)$$

These examples lead us directly to the topic of this book: the use of the moire effect in metrology. The three fields of applications to be discussed in the following chapters, shadow moire, strain analysis, and moire deflectometry, are all based on the idea that a certain disturbance is introduced into a test system. One of the pair of gratings, the sample grating, is distorted accordingly and the distortion is magnified in the moire pattern. From the fringe shift one can quantitatively deduce the system's disturbance.

In shadow moire (Chapter 4), a grating is projected onto an uneven sample (like a relief map). The shadow cast is distorted due to the differences in height and this interferes with the same reference grating to yield a contour map of the surface. In strain analysis

(Chapter 5), a grating is attached to a flat surface element to be tested. The grating is deformed due to stress applied to the sample and this interferes with a reference grating to yield a distorted moire pattern. In moire deflectometry (Chapter 6), the rays deflected by a sample under test, either a specular surface or a phase object, distort the shadow of the sample grating through which they pass. The larger the distance between the sample grating and the reference grating, the greater the distortion of the fringe pattern formed by superposition of the reference pattern with the shadow of the sample grating.

REFERENCES

1. C. Huygens, *Traite de la Lumière*, Leiden, 1690.
2. J. C. Maxwell, *Electricity and Magnetism*, Oxford, 1873.
3. T. Young, *Phil. Trans. Roy. Soc. London xcii*, **12**, 387 (1802).
4. A. A. Michelson, *Phil. Mag. (5)* **13**, 236 (1882).
5. Suggested reading: E. Hecht and A. Zajac, *Optics*, Addison-Wesley, Reading, Mass., 1974, Chaps. 2 and 3.
6. M. Born and E. Wolf, *Principles of Optics*, Pergamon, Oxford, 1970, Chap. 1.
7. J. B. Allen and D. M. Meadows, Removal of Unwanted Pattern from Moire Contour Maps by Grid Translation Techniques, *Appl. Opt.* **10**, 210 (1971).

SYMBOLS

A	amplitude	ε	constant phase term
c	speed of light	λ	wavelength
I	flux energy	δ	phase difference (interferometry)
$\mathbf{k}$	wave vector, $2\pi/\lambda$		
n	refractive index	Ψ	complex wave
p	grating pitch	α	$\mathbf{kr} + \varepsilon$
$\mathbf{r}$	position vector	ϕ	rotation angle
t	time	ω	angular frequency $2\pi v/\lambda$
T	transmittance		
v	velocity		

2

Techniques Used in Moire Metrology

Although this book intends to provide an analogy between interferometric methods applying electromagnetic waves and moire methods
applying gratings, which can be regarded as man-made spatial
waves, a brief review of the techniques used in the field is now
appropriate. We will discuss both those techniques that are unique to
moire and others that are general in optics. The first section is
devoted to fringe quality improvement and to fringe readout techniques. The second section deals with the extraction of different
types of information from the fringes.

2.1. FRINGE QUALITY IMPROVEMENT AND READOUT TECHNIQUES

2.1.1. Grating's Stripes Multiplication and Addition

Grating's stripes multiplication is the preferred way of obtaining
moire fringes. The fringes result from multiplication of the transmission functions of the two gratings and they occur when the two
gratings are superimposed (i.e., in shadow moire, deflectometry, or
strain analysis). Since all the information exists in the distorted
grating, dense gratings should be used to achieve high accuracy.
Stripes multiplication reduces the number of fringes without any loss
in the accuracy. Most viewing devices (camera, eye, etc.) are unable
to resolve the dense pattern of the grating, but can easily resolve the
lower density moire fringes.

A setup for stripes addition, as was suggested by Hovanessian and
Hung [1], is demonstrated in Fig. 2.1. The moire pattern is the sum
of two shadows and is therefore additive. This setup enables much

19

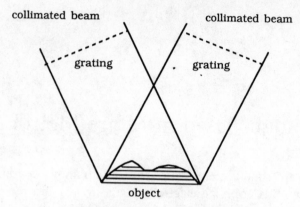

Figure 2.1.   Stripes addition setup.

more sensitivity and flexibility than the regular shadow moire setup (discussed in Chapter 4). Moreover, if the projected grating is formed by interferometric means, such as the straight fringe pattern formed by a Michelson interferometer, there is only the diffraction limit on the density of the stripes projected on the object, and no limit on the depth of field of the projected shadow of the grating. Additive moire fringes can be observed only if we are able to resolve the original gratings' stripes. As in double exposure moire, which is another example of fringe addition, averaging over the grating pitch will destroy the contrast. Although these techniques are extremely useful for mapping contour differences, the requirement of resolving the gratings' stripes dramatically limits the measurement accuracy. The advantage of averaging over the grating pitch, which is allowed only in stripes multiplication techniques, is in reducing the fine structure of the fringe pattern and increasing SNR. In shadow moire, Allen and Meadows [2] suggested moving the grating during exposure to average out the fine structure. In moire deflectometry likewise, Keren [3] suggested moving the two deflectometer gratings in unison to obtain perfect fringes.

2.1.2. Spatial Filtering

The gratings' stripes are of a much higher density than the moire fringes. Therefore, focusing the fringe pattern and using a low pass spatial filter removes the high spatial frequencies and leaves only the moire pattern. For theoretical aspects, see Goodman [4].

2.1.3.   Phase Shift Fringe Readout Method

The phase shift method is based on the mathematical analysis of three patterns shifted by $\frac{1}{3}$ fringe. In its application to moire deflectometry one of the gratings is moved by $\frac{1}{3}$ of its pitch and in shadow moire the projection angle is slightly altered. The idea of the phase shift technique is as follows: The fringe pattern along one axis generally can be described as

$$I(x) = A + B \sin(kx + \varphi). \tag{2.1}$$

If we know the spatial frequency k and we shift the phase three times, we obtain the three equations which are required to find the unknown phase φ and the constants A and B for each point. This sinusoidal equation should be slightly altered for the triangular-shaped moire fringes.

2.1.4.   Electronic Heterodyne Readout Technique

Electronic heterodyne is a more accurate phase shifting technique than phase shift, but for a single point. The idea is common in FM radio receivers. Mathematically speaking the signal of Eq. (2.1) is multiplied by the function $\sin kx$ and averaged over one period by moving the fringe pattern:

$$I_1 = \frac{k}{2\pi} \int_{-\pi/k}^{\pi/k} \sin kx \left[A + B \sin(kx + \varphi) \right] dx. \tag{2.2}$$

We neglect the dc term A because it vanishes when multiplied by a sine wave and averaged over one period. Equation (2.2) can be rewritten as

$$I_1 = \frac{Bk}{4\pi} \int_{-\pi/k}^{\pi/k} \left[\cos \varphi - \cos(2kx + \varphi) \right] dx \tag{2.3}$$

and we obtain

$$2I_1 = B \cos \varphi. \tag{2.4}$$

Similarly we multiply the signal of Eq. (2.1) by $\cos kx$, and averaging over one period yields

$$2I_2 = B \sin \varphi. \tag{2.5}$$

Dividing the two signals results in the phase

$$\varphi = \arctan(I_2/I_1). \tag{2.6}$$

While the resolution in real time measurement, due to the uncertainty principle, is about $1/10$ of a fringe, phase shift processing yields a resolution of $1/100$ of a fringe, and heterodyne is typically 5 times more accurate than phase shift. This improved resolution is explained by mode statistics as discussed in Chapter 3. Application of electronic heterodyne to moire deflectometry was suggested by Stricker [5].

2.1.5.   Other Gratings

Improving the fringe quality is a major issue in optical metrology, and therefore changing the gratings was a subject of extensive research. Good quality gratings are not easily available. We provide a brief review on what was done in this field.

One way to alter the line shape is by changing the opening ratio of the grids, i.e., the ratio between the dark and transparent stripes. In general, if the opening ratio is $1:n$, the fringes will sharpen at the same ratio [6]. An interesting approach suggested by Bryngdahl [7] was to use polarization gratings, in which the dark and transparent stripes are replaced by linear polarizer stripes, which are layered orthogonally at alternating sequence. The idea is that interference, as in electromagnetic waves, can be imitated more closely by polarizers. The orthogonal polarizations will block light in the same way that negative electric field cancels positive electric field. Such gratings were used to obtain higher quality moire patterns (Fig. 2.2) [8].

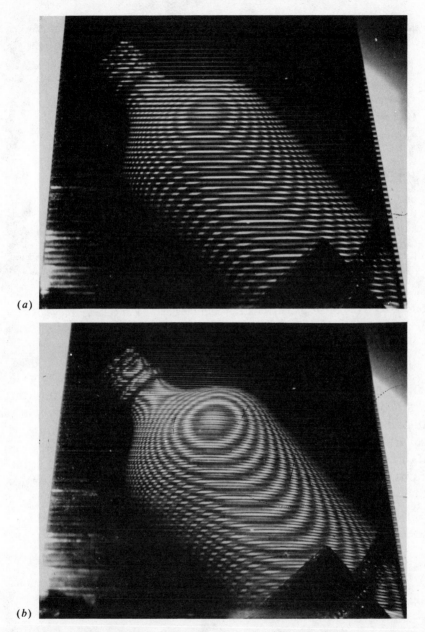

(a)

(b)

Figure 2.2.   (a) Mapping of a diffusive object with a Ronchi ruling. (b) The same object mapped with a polarization grating (linearly polarized light is used).

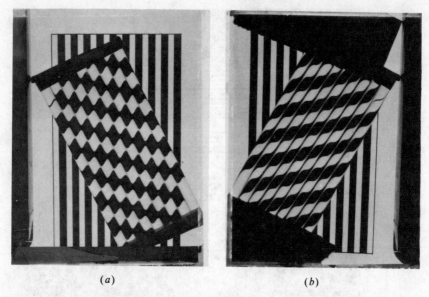

(a)   (b)

Figure 2.3. (a) Moire pattern of a Ronchi ruling and a lenticular grating. (b) At a given angle, perfect Ronchi ruling-type fringes are obtained.

Another approach is to use, as a second grating to a Ronchi ruling, a set of cylindrical lenses with width equal to the pitch of the first Ronchi ruling. This is called *lenticular grating* and is used in toy rulers and three-dimensional (3-D) photographs. The grating is superimposed with a Ronchi ruling at an angle determined by the focal length of the lenticular grating and the distance from the Ronchi ruling. This produces perfect moire fringes, namely, stripes that are identical to those of the Ronchi ruling [9] (Fig. 2.3).

2.2. INFORMATION EXTRACTION FROM DISTORTED GRATINGS

Suppose that we have distorted gratings of the general form

$$y + f_i(x, y) = np, \qquad i = 1, 2, 3, \ldots, \quad n = 0, \pm 1, \pm 2, \ldots \quad (2.7)$$

in which $f_i(x, y)$ is the measured quantity. These gratings can be manipulated to obtain fringe patterns that provide mapping of quantities like $f_i(x, y)$, $f_i(x, y) - f_j(x, y)$, $f_i(x, y) + f_j(x, y)$, or

$$\frac{\partial f_i(x, y)}{\partial x}, \quad \frac{\partial f_i(x, y)}{\partial y}, \quad \frac{\partial^n f_i(x, y)}{\partial x^n}, \quad \frac{\partial^2 f_i(x, y)}{\partial x \, \partial y}, \text{ etc.}$$

We will show the use of some of these techniques in other chapters, and here we provide a rigorous mathematical treatment. n, m, and l can apply any integer value, i.e., n, m, $l = 0, \pm 1, \pm 2, \ldots$. We use different letters only to distinguish between gratings and switch notations in order to save symbols.

2.2.1. Mapping $f(x, y)$

The simplest moire mapping is done by superimposing the undistorted grating with the distorted one to obtain an infinite fringe moire pattern:

$$\begin{aligned}
y + f(x, y) &= np \\
y &= mp \\
\hline
f(x, y) &= lp,
\end{aligned} \qquad \text{where } l = n - m, \qquad (2.8)$$

namely, a contour map of $f(x, y)$ incremented by p. Another solution is the very dense grating

$$2y + f(x, y) = lp, \quad \text{where } l = n + m. \qquad (2.9)$$

Our eye resolves the low density grating and overlooks the high density one. The fringes of Eq. (2.9) can cause problems only when the contours of Eq. (2.8) are dense enough to be comparable to them. Figure 2.4a shows the distorted grating formed by projection of a grating on a human body (see Chapter 4 for details) and Figure 2.4b shows the moire pattern obtained by interference with a Ronchi ruling.

A finite fringe map is obtained by rotating the reference grating by $-\theta/2$ or $\theta/2$ and the distorted grating (2.7) by an angle $\theta/2$ or $-\theta/2$:

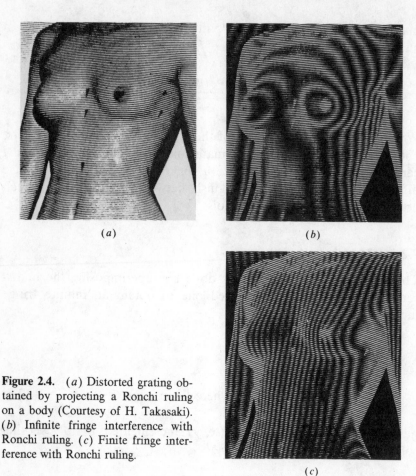

Figure 2.4. (*a*) Distorted grating obtained by projecting a Ronchi ruling on a body (Courtesy of H. Takasaki). (*b*) Infinite fringe interference with Ronchi ruling. (*c*) Finite fringe interference with Ronchi ruling.

$$[y + f(x, y)]\cos(\theta/2) = x \sin(\theta/2) + np,$$
$$y \cos(\theta/2) = -x \sin(\theta/2) + mp. \tag{2.10}$$

By superimposing them we obtain

$$f(x, y)\cos(\theta/2) = 2x \sin(\theta/2) + lp, \qquad l = n - m. \tag{2.11}$$

Considering only small angles where $\cos(\theta/2) \cong 1$ and $\sin(\theta/2) \cong \theta/2$, Eq. (2.11) becomes

$$f(x, y) = x\theta + lp \tag{2.12}$$

or

$$x - \frac{f(x, y)}{\theta} = l\left(\frac{p}{\theta}\right). \tag{2.13}$$

Equation (2.13) describes a grating perpendicular to the bisector of the original gratings, with pitch and distortion function magnified by $1/\theta$. An example of a finite fringe pattern is seen in Fig. 2.4c. Grating (2.13) can produce a moire pattern of $f(x, y)$ by superimposing it with a coarse grating of pitch p/θ,

$$x = n\left(\frac{p}{\theta}\right), \tag{2.14}$$

and obtaining

$$f(x, y) = mp, \qquad m = l - n. \tag{2.15}$$

This is exactly the same result that was obtained with the original gratings [Eq. (2.8)].

It is important to distinguish between two types of moire patterns. In deflectometry and shadow moire, the frame of reference is determined by the object. The coordinates of the function f are not transformed by rotating the gratings. In strain analysis the test object is a distorted grating and rotating it causes the coordinates of f to transform. Nevertheless, in the small angle approximation the result will hardly change because Eq. (2.10) can be rewritten as

$$y + f(x, y) = np,$$
$$y \cos \theta = -x \sin \theta + mp. \tag{2.10'}$$

In other words, rotating the undistorted grating, rather than the mutual rotation of the two gratings, is less accurate because the approximation $2 \sin(\theta/2) \sim \theta$ is better than $\sin \theta \sim \theta$.

2.2.2. Mapping of Differences

When the difference between two functions rather than the absolute value is desired, the two distorted gratings are superimposed to form a moire pattern:

$$y + f_1(x, y) = np,$$
$$y + f_2(x, y) = mp, \tag{2.16}$$

which results in

$$f_1(x, y) - f_2(x, y) = lp. \tag{2.17}$$

This is a contour map of the difference between the two functions which is incremented by p.

2.2.3. Mapping the Sum [10]

This is not as simple because producing a conjugate grating is necessary. In interferometry, in order to sum two functions, one is given by a wave $\exp\{i\varphi_1(x, y)\}$ and the other by a conjugate wave of the form $\exp\{-i\varphi_2(x, y)\}$. In moire, two distorted gratings are required:

$$y + f_1(x, y) = np \tag{2.18}$$

and

$$y - f_2(x, y) = mp. \tag{2.19}$$

To obtain *conjugate gratings* we use the finite fringe technique. The first grating is the finite fringe map obtained in Eq. (2.13):

$$x - \frac{f_1(x, y)}{\theta} = n\left(\frac{p}{\theta}\right). \tag{2.20}$$

The second grating is produced by reversing the angle $\theta/2$ of the original grating versus the reference grating

$$[y + f_2(x, y)]\cos(-\theta/2) = x \sin(-\theta/2) + np,$$
$$y \cos(-\theta/2) = -x \sin(-\theta/2) + mp, \tag{2.21}$$

and we obtain

$$x + \frac{f_2(x, y)}{\theta} = l\left(\frac{p}{\theta}\right), \qquad l = n - m. \tag{2.22}$$

The sum of the finite fringe maps (2.20) and (2.22) is the contour map

$$f_1(x, y) + f_2(x, y) = mp. \tag{2.23}$$

2.2.4. Multiplication by a Factor

In order to multiply $f(x)$ by a given factor M we use the beats phenomenon described in the previous chapter, namely, we use the moire pattern formed by the two gratings

$$y + f(x, y) = np_1, \tag{2.24}$$

$$y = mp_2, \tag{2.25}$$

where $p_1 \neq p_2$. To solve this set of equations we substitute n and m and obtain

$$l = m - n = \frac{y + f(x, y)}{p_1} - \frac{y}{p_2}. \tag{2.26}$$

Introducing the beats pitch from Chapter 1,

$$\frac{1}{p} = \frac{1}{p_1} - \frac{1}{p_2}, \tag{2.27}$$

Figure 2.5.   Moire pattern of beats obtained with a sinusoidal grating and a Ronchi ruling of different pitches. The beats pattern is a magnified sinusoidal grating.

we have,

$$y + Mf(x, y) = lp, \tag{2.28}$$

where $M = p/p_1$. This result is demonstrated in Fig. 2.5 where a sinusoidal grating is superimposed with a Ronchi ruling of different pitch. The resulting moire pattern is identical to the original grating, but the pitch is that of the beats and the sine wave is magnified.

2.2.5. Weighted Sums and Differences

In a similar way we can subtract two weighted functions f_1 and f_2 so that we obtain $M_1 f_1 - M_2 f_2$. This is done by superimposing the two gratings

$$y + f_1(x, y) = np_1, \tag{2.29}$$

$$y + f_2(x, y) = mp_2. \tag{2.30}$$

The moire pattern obtained is

$$y + \frac{p}{p_1} f_1(x, y) - \frac{p}{p_2} f_2(x, y) = lp. \tag{2.31}$$

A contour map of the weighted difference can be obtained by superimposing this pattern with a Ronchi ruling with pitch equal to the beats period. By analogy, a contour map of the weighted sum can be obtained.

2.2.6. Partial Derivatives

These are obtained by combining the grating described by Eq. (2.7) with its replica shifted in a given direction δr. Assuming that,

$$\delta r = a\,\delta x + b\,\delta y \quad \text{where } (a^2 + b^2)^{1/2} = 1, \tag{2.32}$$

then

$$y + b\,\delta y + f(x + a\,\delta x, y + b\,\delta y) = np,$$
$$y + f(x, y) = mp. \tag{2.33}$$

The result is

$$f(x + a\,\delta x, y + b\,\delta y) + b\,\delta y - f(x, y) = lp, \qquad l = n - m, \tag{2.34}$$

where $b\,\delta y$ is a phase constant which will be omitted for simplification. We divide Eq. (2.34) by δr and obtain

$$\frac{f(x + a\,\delta x, y + b\,\delta y) - f(x, y)}{\delta r} \cong \frac{\partial f(x, y)}{\partial r} \cong \frac{lp}{\delta r}, \tag{2.35}$$

which describes a contour map of the partial derivative incremented by $p/\delta r$. Figure 2.6 presents partial derivative fringes of a distorted grating. To obtain the derivative we can also shift the contour map of the function,

$$f(x, y) = np,$$
$$f(x + \delta x, y) = mp, \tag{2.36}$$

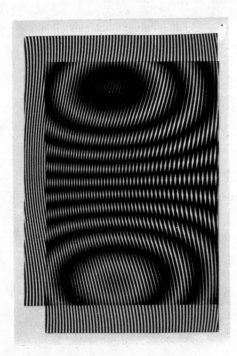

Figure 2.6. Partial derivative fringes obtained by superposition of shifted gratings.

which yields

$$\frac{\partial f}{\partial x} \cong \frac{lp}{\delta x}, \qquad l = m - n. \qquad (2.37)$$

This technique may result in a pattern with very few fringes or none at all. To overcome this problem we use the finite fringe mode. In the finite fringe mode the number of fringes can be controlled, regardless of the magnitude of the function $f(x, y)$. There is, however, a limit to the number of fringes because a relatively small θ should be used, both to keep the fringes smooth and due to the small angle approximation used in the derivations. To analyze the error due to the small angle approximation we write Eq. (2.13) exactly:

$$x + \frac{f(x, y)}{2\tan(\theta/2)} = \frac{lp}{2\sin(\theta/2)}. \qquad (2.38)$$

By expanding the trigonometric functions to a Taylor series,

$$2\sin\left(\frac{\theta}{2}\right) = \theta - \frac{\theta^3}{24} + \cdots, \qquad (2.39)$$

$$2\tan\left(\frac{\theta}{2}\right) = \theta + \frac{\theta^3}{12} + \cdots, \qquad (2.40)$$

we see that in order to meet an accuracy better than 1%, θ should be smaller than 0.28 rad or 16°. This angle will reduce the number of fringes by ~ 75%.

To obtain higher-order derivatives, we form a finite fringe pattern of the partial derivative. This is done by rotating and shifting the grating of Eq. (2.7), namely,

$$[y + f(x, y)]\cos(\theta/2) = x(\sin\theta/2) + np,$$
$$[y + b\,\delta y + f(x + a\,\delta x, y + b\,\delta y)]\cos(\theta/2)$$
$$= -(x + a\,\delta x)\sin(\theta/2) + mp, \qquad (2.41)$$

which results in

$$b \, \delta y + f(x + a \, \delta x, y + b \, \delta y) - f(x, y) \cong -x\theta + \frac{a \, \delta x \, \theta}{2} + lp,$$

$$l = m - n. \quad (2.42)$$

Neglecting the constant phase terms $b \, \delta y$ and $a \, \delta x \, \theta/2$ and dividing by θ, we obtain

$$x + \frac{\delta r}{\theta} \frac{\partial f(x, y)}{\partial r} \cong l\left(\frac{p}{\theta}\right), \quad (2.43)$$

which describes a grating of pitch p/θ in which the partial derivative of f is mapped. To obtain the second derivative one can shift the grating of Eq. (2.43) in the direction

$$\delta r_1 = a_1 \, \delta x + b_1 \, \delta y \quad (2.44)$$

and combine it with its replica:

$$x + \frac{\delta r}{\theta} \frac{\partial f(x, y)}{\partial r} = l\frac{p}{\theta}, \quad (2.45)$$

$$x + a_1 \, \delta x + \frac{\delta r}{\theta} \frac{\partial f(x + a_1 \, \delta x, y + b_1 \, \delta y)}{\partial r} = m\frac{p}{\theta}. \quad (2.46)$$

The moire pattern is

$$a_1 \, \delta x + \frac{\delta r}{\theta} \left[\frac{\partial f(x + a_1 \, \delta x, y + b_1 \, \delta y)}{\partial r} - \frac{\partial f(x, y)}{\partial r} \right] = n\frac{p}{\theta},$$

$$n = m - l. \quad (2.47)$$

As before we neglect the constant phase term and obtain

$$\frac{\partial f(x + a_1 \, \delta x, y + b_1 \, \delta y)}{\partial r} - \frac{\partial f(x, y)}{\partial r} = \frac{np}{\delta r} \quad (2.48)$$

or

$$\frac{\partial^2 f(x, y)}{\partial r \, \partial r_1} \cong \frac{np}{\delta r \, \delta r_1}, \tag{2.49}$$

which is a contour map of the second derivative incremented by $p/\delta r \, \delta r_1$. This operation can be repeated to obtain higher derivatives by producing the finite fringe pattern of Eq. (2.49).

2.2.7. Self-Healing

A nice property of the moire pattern is its immunity to the gratings' noise function $n(x, y)$, when the two superimposed gratings carry the same noise. This is shown as

$$y + n(x, y) = np,$$
$$y + n(x, y) = mp \tag{2.50}$$

and results in infinite fringe.

Figure 2.7. Straight moire fringes obtained with two identical distorted gratings.

This property may be used to observe a signal $f(x, y)$ on a noisy grating, namely,

$$y + n(x, y) + f(x, y) = np, \tag{2.51}$$
$$y + n(x, y) = mp \tag{2.52}$$

yield

$$f(x, y) = lp, \qquad l = n - m. \tag{2.53}$$

Note that the noise function $n(x, y)$ was eliminated. Even for finite fringe patterns this approximation may be useful if the noise function has a small tangential derivative. Figure 2.7 shows straight fringes obtained from two distorted gratings.

2.2.8. Grating and Fringe Pattern Phase

So far we have combined the two gratings so that their stripes overlap without a phase shift. This is a special case of the general situation in which a certain phase Ψ exists between the two gratings:

$$\Psi = \frac{2\pi y_0}{p}. \tag{2.54}$$

Adding the relative phase to the two gratings of Eq. (2.10) yields

$$[y + f(x, y)]\cos(\theta/2) = x \sin(\theta/2) + np,$$
$$y \cos(\theta/2) = -x \sin(\theta/2) + mp + \Psi p/2\pi, \tag{2.55}$$

which results in

$$x + \frac{f(x, y)}{\theta} \cong l\left(\frac{p}{\theta}\right) + \left(\frac{\Psi}{\theta}\right)\frac{p}{2\pi}, \qquad l = n - m. \tag{2.56}$$

Note that the phase is multiplied by $1/\theta$. This is a mathematical explanation to the one period fringe shift of the moire pattern that results from shifting the gratings a distance p with respect to each other.

2.2.9. Spatial Correlation of Two Moire Patterns [11]

The spatial optical correlator is based on superposition of the moire fringe patterns of two objects to be correlated. It is used for comparing, or for the convolution of two moire patterns. The spatial correlation function $C(r)$ of the patterns, represented by the functions $f(r)$ and $g(r)$ which are sheared by r, is given by

$$C(r) = \int_{-\infty}^{\infty} f(r')g(r' + r) \, dr'. \qquad (2.57)$$

The measurement requires infinite fringe patterns in which the phase shift is between 0 and π (i.e., the transmittance should be proportional to the relative shift between the gratings as in the case of Ronchi rulings; see Chapter 1). The transmittance at each point is proportional to the product of the sheared functions, and integration is obtained with a collecting lens. The spatial correlation as a function of the shear is measured by translating one moire pattern with respect to the other. Convolution is obtained in the same manner by inverting one of the transparancies f or g.

2.2.10. Radial and Circular Gratings

Although gratings other than Ronchi rulings can be used in moire analysis, their impact is limited. We mention the radial and circular gratings because of their benefit in symmetry problems. Although many authors have described their moire patterns, the only practical moire measurement with circular gratings, to our knowledge, is the analysis of the shift of two beams with respect to each other [12, 13]. The circular gratings are represented in polar coordinates as,

$$r = np. \qquad (2.58)$$

This is analogous to the Cartesian coordinate representation of the grating of Eq. (2.7). The distorted grating will be

$$r + f(r, \theta) = mp. \qquad (2.59)$$

The result is analogous to that of Ronchi rulings, namely,

$$f(r, \theta) = lp, \qquad l = m - n. \qquad (2.60)$$

Similarly the radial grating is given by

$$\theta = n\theta_p, \qquad (2.61)$$

where $2\pi/\theta_p$ is the integer k and $n = 0, 1, 2, \ldots, k$.

REFERENCES

1. J. Der Hovanessian and Y. Y. Hung, Moire Contour Sum, Contour Difference and Vibration Analysis of Arbitrary Objects, *Appl. Optics* **10**, 2734 (1971).

2. J. B. Allen and D. M. Meadows, Removal of Unwanted Patterns from Moire Contour Maps by Grid Translation Techniques, *Appl. Opt.* **10**, 210 (1971).

3. E. Keren, Immunity to Shock and Vibration in Moire Deflectometry, *Appl. Opt.* **24**, 3028 (1985).

4. J. W. Goodman, *Introduction to Fourier Optics*, McGraw-Hill, New York, 1968, Chap. 8.

5. J. Stricker, Diffraction Effects and Special Advantages in Electronic Heterodyne Moire Deflectometry, *Appl. Opt.* **25**, 895 (1986).

6. See, for example, H. Takasaki, Moire Topography from its Birth to Practical Application, *Optics and Lasers in Engineering* **3**, 3 (1982).

7. O. Bryngdahl, Polarization-Grating Moire, *J. Opt. Soc. Am.* **6**, 839 (1982).

8. A. Livnat and O. Kafri, Polarization Gratings Phase Rulings and Moire Analysis, *Opt. Lett.* **6**, 266 (1981).

9. A. Livnat and O. Kafri, Moire Pattern of Linear Grid with Lenticular Gratings, *Opt. Lett.* **7**, 253 (1982).

10. A. Livnat and O. Kafri, Fringe Addition in Moire Analysis, *Appl. Opt.* **22**, 3103 (1983).

11. O, Kafri, T. Chin, and D. F. Heller, Moire Optical Spatial Correlator, *Opt. Lett.* **9**, 481–3 (1984).

12. D. F. Heller, O. Kafri, and J. Krasinski, Direct Birefringence Measurements Using Moire Ray Deflection Techniques, *Appl. Opt.* **24**, 3037 (1985).

13. I. Glatt and O. Kafri, Beam Direction Determination by Moire Deflectometry Using Circular Gratings, *Appl. Opt.* **26**, 4051 (1987).

SYMBOLS

I	Intensity		
C	correlation function	$\delta \mathbf{r}$	shear
k	spatial frequency	θ	angle between gratings
n, m, l	indicial equation	φ	phase
	constants	Ψ	phase
M	multiplication factor		
p	grating pitch		

3

Limitations on Accuracy Due to the Use of Light

As we saw in the previous chapter, a great deal of information can be extracted by moire analysis of a distorted grating. The distorted grating is formed either by projection of a reference grating onto a test object (as in shadow moire) or by projecting a reference grating with a distorted beam (as in moire deflectometry). Therefore, the limiting resolution and accuracy is determined by the properties of light. Moreover, as will be shown in Chapters 6 and 7, we can learn a great deal about the properties of light from moire measurements. This is similar to the way interferometry gives us information on both properties of light and matter. This chapter deals with light properties and the limits on measurements done with light.

3.1. INTRODUCTION

Light is an electromagnetic phenomenon. The rigorous treatment of electromagnetic wave propagation is derived from the electromagnetic equations formulated by Maxwell (1831–1879) and from quantum mechanics. However, when the wavelength of light is small compared with the dimensions of the measuring apparatus, an approximate approach, known as geometrical or ray optics, can be applied to solve most problems concerning light propagation.

An analogy between optics and mechanics has been drawn by Marcuse in his book *Light Transmission Optics* [1]. Geometrical optics is analogous to the classical mechanics of point particles, whereas wave optics can be compared to quantum mechanics. Thus, by applying the general Hamiltonian formulation of classical mechanics to optics, one can deduce the basic ray equation that is conventionally derived from the Fermat principle of least time [2] (as well as other fundamental laws in optics such as the law of reflection and Snell's law).

In this chapter, the properties of light are reviewed briefly, and their effects on the accuracy of optical measurements are studied. We focus our discussion on the problem of beam divergence, which is detrimental to ray deflection analysis. Beam divergence, as will be shown, is imminent with the wave nature of light, as manifest by the diffraction phenomenon. It is further increased on passing through diffusive media, an irreversible process which leads to the reduction of beam quality. The term *beam divergence* in the present context, refers exclusively to irreversible beam divergence, which in contrast to the effect of a diverging lens, cannot be reversed by the use of passive optical elements.

Traditionally, optical objects are classified in three groups: transparent or phase objects, specular objects, and diffusive objects. The first two categories include objects that redirect the beams without degradation of the beam quality. They may distort the wavefront upon transmission or reflection, but the directional character of the beam is conserved. Diffusive objects, on the other hand, scatter the light rays into multiple directions, thereby increasing the divergence. The classification is indeed artificial and stems probably from the different approaches to the analysis of the two types of optical objects, namely, interferometric methods for specular and wavefront distorting objects and holography or shadow moire for diffusive objects. Real optical objects are always diffusive to a certain extent and hence reduce the beam quality. In this chapter, beam quality is expressed in terms of radiation modes and is related to the concept of light coherence.

3.2. PROPERTIES OF LIGHT

Light is an electromagnetic disturbance propagating in space at a velocity c/n, where c is the velocity of light in vacuum and n is the index of refraction, a characteristic property of the medium (n

equals unity in free space). The electric field of light, denoted by the vector **E**, is a solution of the differential wave equation

$$\nabla^2 \mathbf{E} = \frac{n^2}{c^2} \frac{\partial^2 \mathbf{E}}{\partial t^2}$$

$$\left(\nabla^2 \equiv \frac{\partial^2}{\partial x^2} + \frac{\partial^2}{\partial y^2} + \frac{\partial^2}{\partial z^2} \text{ is the Laplacian operator} \right) \quad (3.1)$$

derived from the Maxwell equations in free space [3]. Any wave function that is a solution of the scalar differential wave equation constitutes a component of the electromagnetic field. Thus, light can be described as a superposition of an infinite number of plane monochromatic waves of the general form $A(\omega)e^{i\omega t}$, where $A(\omega)$ is the amplitude of the harmonic wave oscillating at angular frequency ω and the exponential term shows the phase change in time (more accurately, the motion is given by a sine or cosine function, i.e., the wave should be described as Re[$A(\omega)e^{i\omega t}$]). The electric field $E(t)$ at a time t can be given by the integral

$$E(t) = \int_{-\infty}^{\infty} A(\omega)e^{-i\omega t}\, d\omega, \quad (3.2)$$

which states that $E(t) = \mathscr{F}\{A(\omega)\}$. In words, the electric field is the Fourier transform of the amplitude $A(\omega)$. Therefore, $A(\omega)$ is the inverse Fourier transform of the electric field, namely,

$$A(\omega) = 2\pi \int_{-\infty}^{\infty} E(t)e^{i\omega t}\, dt. \quad (3.3)$$

The functions $A(\omega)$ and $E(t)$ are a Fourier transform pair, a relationship whose physical meaning is discussed in the subsequent text.

To understand the random nature of light it is useful to regard nearly monochromatic, i.e., quasimonochromatic light (purely monochromatic light sources do not exist), as a series of randomly phased wave trains of finite duration, oscillating at a characteristic frequency. Fourier decomposition of such a wave train results in an infinite series of contributions of all frequencies, centered around the characteristic frequency. The longer the wave train, the narrower the frequency distribution. The representative bandwidth $\delta\nu$ of the Fourier transform is arbitrarily defined as the frequency gap between the two half height points of the frequency distribution. The reciprocal of $\delta\nu$ is known as the *coherence time*, δt ($\delta t = 1/\delta\nu$). Within the interval δt, light behaves perfectly monochromatic, and the phase at any point in space can be reasonably predicted.

Another useful term derived from the coherence time is the *coherence length*, $\delta x = c\,\delta t$, which can be regarded simply as the length of a coherent wave train. The coherence length is a measure of the spectral purity of light. Purely monochromatic light has an infinite coherence length, whereas in incoherent light (i.e., light with a very small coherence length), the phase information is lost within a short distance. We can deduce that the coherence property of light is closely connected with the autocorrelation function of the wave functions. To demonstrate, in a Michelson interferometer where two beams of the same phase, originating from the same light source, are superimposed after traveling different optical paths, the intensity, or the contrast, of the interference pattern is determined by the degree of coherence of the light. Highly coherent light will yield a sharp fringe pattern and a poorly coherent source will result in a blurred pattern [3]. Therefore, to minimize the optical path difference be-

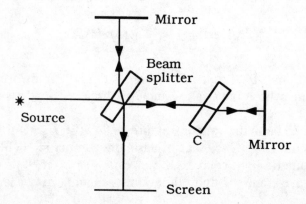

Figure 3.1.   Coherence length measurement with a Michelson interferometer.

tween the two arms of the interferometer, a compensator plate C is inserted in one of the arms (see Fig 3.1).

The degree of coherence of a point source $\gamma_{11}(\tau)$ (τ denotes the time gap over which the correlation is performed) is defined as the normalized dimensionless autocorrelation function, namely,

$$\gamma_{11}(\tau) = \frac{\Gamma_{11}(\tau)}{\langle |E(t)|^2 \rangle}. \tag{3.4}$$

The brackets are symbols for time average and $\Gamma(\tau)$ is the autocorrelation function given by

$$\Gamma_{11} = \langle E(t+\tau)E^*(t) \rangle. \tag{3.5}$$

The absolute value of $\gamma_{11}(\tau)$ may vary between zero (total incoherence) and unity (full coherence). Light with intermediate $\gamma(\tau)$ values is said to be partially coherent. The modulus of the complex degree of temporal coherence $\gamma_{11}(\tau)$, which relates to phase fluctuations in time, is identical to the visibility. The visibility is also called the modulation M of the fringe pattern, formulated by Michelson as

$$M = \frac{I_{max} - I_{min}}{I_{max} + I_{min}}, \tag{3.6}$$

where I_{max} and I_{min} are the intensities of the maximum and adjacent minimum in the fringe array. Thus, the modulus of $\gamma_{11}(\tau)$ can be directly calculated from the contrast of the interference pattern.

3.3. LIGHT COHERENCE AND RADIATION MODES

We now introduce a rather useful measure for light quality involving radiation mode statistics and based on the Heisenberg uncertainty principle. For a pulse of duration T, the uncertainty in time of occurrence can also be T. If $\delta\nu$ is the frequency bandwidth, defined as the reciprocal of the coherence time, then according to the uncertainty principle [4],

$$\delta\nu\, T \geq 1. \tag{3.7}$$

The ratio of the pulse duration to the coherence time δt is m, namely,

$$m = T/\delta t. \tag{3.8}$$

Thus, m is the number of coherence lengths existing in a pulse. We define it as the number of longitudinal modes of radiation in the light pulse.

The total number of electromagnetic modes of unpolarized light N, having frequencies in the range between ν and $\nu + \delta\nu$ and enclosed in a volume V, is found from Maxwell's equations for a particle confined in a box [5]:

$$N = \frac{8\pi\nu^2\,\delta\nu\,V}{c^3}. \tag{3.9}$$

The fraction of modes contained within a solid angle $\delta\Omega$ would thus be

$$N(\Omega) = \frac{2\nu^2}{c^3}V\,\delta\nu\,\delta\Omega. \tag{3.10}$$

V actually denotes the volume from which the light originally emerged.

For simplicity, let us assume that the beam in question is a pulse of duration T with a square cross section a^2 at the waist. The initial volume is, therefore, a^2cT and

$$N(\Omega) = \frac{2\nu^2}{c^2}\,\delta\nu\,\delta\Omega\,a^2T \equiv 2\,\delta\nu\,\delta\Omega\,T\frac{a^2}{\lambda^2}. \tag{3.11}$$

Since $\delta\nu\,T$ was already defined as m, the number of longitudinal modes of radiation in a pulse (i.e., the number of modes along the direction of propagation), we obtain,

$$N(\Omega) = 2m\,\delta\Omega\left(\frac{\lambda}{a}\right)^2. \tag{3.12}$$

The term λ/a is approximately the far field (i.e., Fraunhofer) diffraction-limited divergence in one dimension by a slit of aperture a [6]. If z is the distance from the effective aperture, then $z\lambda/a$ is the width of the central diffraction lobe, defined as the distance between the first zero values of the diffraction pattern. Therefore, $(\lambda/a)^2$ is the solid angle of the two-dimensional diffraction-limited divergence of an ideally collimated beam with a square aperture. The real divergence observed $\delta\Omega$ is expected to be higher than the diffraction limited. The ratio $n = \delta\Omega/(\lambda/a)^2$ is referred to as the number of transverse modes of radiation. The total number of radiative modes N, as defined before, is given by the product $2nm$. The number 2 stands for unpolarized light where two quantum states exist. For polarized light, $N = nm$.

In the same manner that m, the number of longitudinal modes, determines the spectral quality or temporal coherence of light, n relates to the beam quality. It determines the degree of beam collimation or the extent to which a beam can be focused by a converging system, namely, the minimum spot size. To demonstrate, suppose we use a lens of focal length f and effective aperture a (f number $= f/a$). The diffraction-limited spot size of such a system is

$$x = \frac{\lambda f}{a}. \tag{3.13}$$

Technically, it is difficult to manufacture high quality lenses with f number < 1, and therefore it is quite impossible to focus a beam to a spot smaller than λ. Similarly, a beam with n transverse modes cannot be focused better than a spot size $n\lambda$. The number of transverse modes defines the spatial coherence in the same way as the number of longitudinal modes defines the temporal coherence. If we focus a light beam to a spot size of $n\lambda$, we can relate to each distance λ as an independent one-dimensional coherent source. Therefore, there is no phase relation between the sources. If such a correlation existed, the beam could have been further focused to a smaller spot size. There is an intimate relation between the coherence function, the fringe modulation, and the number of radition modes. The time correlation function Γ_{11} is the coherence length distribution and its width is, therefore, $1/\delta\nu$. The number of longitudinal modes is the pulse duration divided by the width of $\Gamma(t)$. The number of transverse modes is the beam diameter divided by the width of spatial coherence function $\Gamma_{11}(x)$. The fringe modulation can reveal m or n depending if the correlation is done in space or time.

3.4. LIMITATIONS ON SPATIAL RESOLUTION—THE UNCERTAINTY PRINCIPLE

As shown in the previous section, it is practically impossible to focus a beam to a smaller spot size than the wavelength λ. To increase the spatial resolution one could employ shorter wavelength sources. But as will be shown, an increase in the spatial resolution results in a decrease in the angular resolution. This interrelation between the angular and spatial resolution is an inevitable result of the Heisenberg momentum-space uncertainty principle, which is demonstrated in optics by the diffraction phenomenon. The smaller the aperture of a collimated beam, the larger the diffraction angle.

The uncertainty principle, resulting from the commutation relations between the quantum mechanical operators related to the coordinate and momentum variables, decrees that the product of the uncertainties in the simultaneous determination of a particle's location and momentum cannot be as small as desired, but rather has a finite bound. Define the uncertainty in determining the location δx as the average deviation from the expectation value $\langle x \rangle$, i.e., $\delta x = \langle (x - \langle x \rangle)^2 \rangle^{1/2}$ and the uncertainty in measuring the momentum as $\delta p_x = \langle (p_x - \langle p_x \rangle)^2 \rangle^{1/2}$. Then the uncertainty principle states that [1]

$$\delta x \, \delta p_x \geq h/2\pi. \tag{3.14}$$

Replacing p_x by the quantum-mechanical equivalent $(h/2\pi)k_x$ (where $|k| = 2\pi/\lambda$ and $\mathbf{k}$ is the propagation vector) and dividing the expression on both sides by p_x we obtain

$$\delta x (\delta p_x/p_x) \geq \lambda/2\pi. \tag{3.15}$$

In the small angle approximation $\delta p_x/p_x$ expresses the uncertainty in ray direction or the angular spread $\delta\phi$. We can therefore write Eq. (3.15) as

$$\delta x\, \delta\phi_x \geq \lambda/2\pi. \tag{3.16}$$

In words, when the spatial resolution required is δx, the angular resolution cannot be better than $\lambda/2\pi\,\delta x$ and vice versa. We can therefore deduce that if the spatial resolution required is no better than the beam aperture, the angular resolution is only limited by the diffraction angle. Furthermore, by invoking the paraxial approximation on the uncertainty principle and stating that a ray deflected off-axis by a small angle ϕ (i.e., $\sin\phi \approx \phi$) will be shifted upon propagating a distance z by $x' \approx z\phi(x' \ll z)$, we can derive the expression for the far field (Fraunhofer) diffraction. This is achieved by analogy to the Fourier transform relation in the time-frequency domain. For a temporally coherent pulse we have obtained the time behavior from the Fourier transform,

$$f(t) = \int_{-\infty}^{\infty} F(\omega)e^{i\omega t}\, d\omega \tag{3.17}$$

and applying the inverse Fourier transformation, the spectrum can be obtained from the time behavior.

Similarly we can deduce the angular distribution from the spatial distribution. From the dimensionless expression relating $\delta\phi$ and δx,

$$\frac{\delta\phi\, \delta x}{\lambda} \sim 1, \tag{3.18}$$

we can write the Fourier transform pair,

$$f(\phi) = \int_{-\infty}^{\infty} F(x)\exp\left(\frac{i2\pi\phi x}{\lambda}\right) dx \tag{3.19}$$

$$F(x) = \frac{1}{\lambda}\int_{-\infty}^{\infty} f(\phi)\exp\left(-\frac{i2\pi\phi x}{\lambda}\right) d\phi.$$

By substituting ϕ with x'/z and renormalizing as shown above we obtain,

$$f\left(\frac{x'}{z}\right) = \frac{1}{\sqrt{\lambda z}} \int_{-\infty}^{\infty} F(x)\exp\left(\frac{i2\pi x'x}{\lambda z}\right) dx. \qquad (3.20)$$

Up to a fixed phase factor, Eq. (3.20) is exactly the expression for diffraction in the Fraunhofer approximation [7]. In other words, the field $f(x'/z)$ at a sufficiently long distance z (i.e., $z > x^2/\lambda$) from an aperture is merely the Fourier transform of the field at $z = 0$.

3.5. MODE STATISTICS AS A MEANS TO IMPROVE THE OPTICAL RESOLUTION

The optical resolution attainable in measurement is limited by the quantum nature of light, as expressed in the Heisenberg uncertainty principle. On this basis, is it possible to increase the spatial resolution without increasing the diffraction effect? This question is critical for moire deflectometry where we wish to measure extremely small ray deflections without totally losing the spatial resolution.

To demonstrate the severity of the problem, the diffraction angle of a He–Ne laser beam ($\lambda = 0.63$ μm) of width 1 mm is $\phi = (\lambda/a)$ rad. After propagating 100 m, the beam diverges to a diameter of ~ 0.63 cm.

The solution suggested is based on mode statistics, in a similar manner as averaging over a large number of events reduces the standard deviation by a factor of the square root of the number of events. The number of events is not identical to the number of photons emitted, but to the number of longitudinal modes of radiation within the pulse. This is because all photons belonging to the same quantum state (i.e., mode) are indistinguishable and therefore refer to a single statistical event. The number of longitudinal modes in a pulse m is given by the product of the bandwidth $\delta\nu$ and the duration T. The definitions of $\delta\nu$ and T may be cumbersome because the distribution functions $f(\nu)$ or $f(t)$ are not always simple Gaussian-like functions with a well-defined bandwidth. Nevertheless, it is not always necessary to go back to the coherence function in order to enjoy the simplified treatment of modes. To demonstrate a more complex case, we analyze laser radiation. The spectrum of laser radiation comprises an envelope of the overall transmission $\Gamma(\nu)$ having spectral width $\delta\nu$ (see Fig. 3.2). Under this envelope there are the radiation modes that are separated by $c/2L$ (where L is the cavity length) and having width $\Delta\nu$ [9] determined by the noise character of the laser. The number of longitudinal modes is thus given by

$$m = \delta\nu\, T/m_t, \quad \text{where } m_t = \frac{c/2L}{\Delta\nu}. \tag{3.21}$$

The ratio of the frequency bandwidth actually filled by the laser to the frequency bandwidth of each cavity mode is m_t^{-1}. Substituting m_t in Eq. (3.21) yields the weighted number of longitudinal modes

$$m = \frac{T\,\delta\nu\,\Delta\nu}{c/2L}. \tag{3.22}$$

A *single-mode laser*, which consists by definition of a single cavity mode, i.e., $\delta\nu \approx c/2L$, contains $T\Delta\nu$ longitudinal modes. Only when T equals the coherence time $\Delta\nu^{-1}$ is m equal to unity.

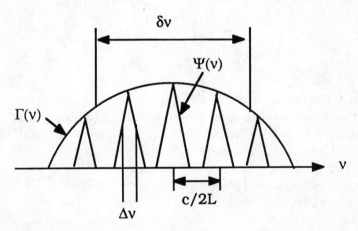

Figure 3.2. Spectrum of a broad band laser. The gain function $\Gamma(\nu)$ has width $\delta\nu$ while the cavity modes $\Psi(\nu)$ have width $\Delta\nu$.

The resolution R in locating the center of a beam consisting of m longitudinal modes is given by

$$R = \delta x'/\sqrt{m} \tag{3.23}$$

according to the central limit theorem [8], namely, an improvement by $m^{-1/2}$ compared to the single-mode resolution. In counting the number of effective modes, only those modes sufficiently populated to be detected, should be considered. Therefore, the emission bandwidth $\delta\nu$ is limited to a frequency range in which the number of photons per mode is above the detection threshold. Therefore, $\delta\nu$ is a function of the sensitivity of the detection system as well as the spectral content of the light source. In any case, the upper limit of $\delta\nu$ is the spectral content of the light source.

Equation (3.22) gives us the relationship between the exposure time and the number of longitudinal modes. The spatial resolution $\delta x'$ at a distance z from the origin is given by

$$\delta x' = z\,\delta\phi. \tag{3.24}$$

Since $\delta\phi \geq \lambda/2\pi\,\delta x$, we can write the inequality for the time required to obtain a given spatial resolution R at a distance z from the origin of a laser light source with diameter δx, namely,

$$t \geq \left(\frac{\lambda z}{2\pi R\,\delta x}\right)^2 \frac{c}{2L} \frac{1}{\delta\nu\,\Delta\nu}. \tag{3.25}$$

To illustrate the application of Eq. (3.25) we provide a numerical example. Suppose that it is required to resolve the center of a Gaussian He–Ne laser beam. The specifications of a typical beam are: the diameter δx, where the power density drops to $1/e^2$ of the maximum at the beam center, is 0.8 mm, $\delta\nu = 2 \times 10^9$ Hz, $\Delta\nu = 10^6$ Hz, and $L = 300$ mm. The exposure time t_{min} required to obtain the resolution R is

$$t_{min} = \left(\frac{\lambda z}{R}\right)^2 (5 \times 10^{-8})\text{ s},$$

where z is given in millimeters. For example, if the R required is 10 λ at $z = 10$ m, then t_{min} is 0.05 s. For $R = \lambda$ an exposure time of 5 sec is required.

3.6. SUMMARY

In this chapter we discussed the limits on accuracy of optical measurements imposed by the properties of light. The wave nature of light leads to the quantum mechanical uncertainty principle, which is manifest by the familiar diffraction phenomenon. We introduced the concept of radiative modes as a quantitative measure of beam quality, and finally discussed the use of mode statistics in improving the spatial resolution without reducing the angular resolution of the measurement.

REFERENCES

1. D. Marcuse, *Light Transmission Optics*, Van Nostrand-Reinhold, Princeton, N.J., 1972.
2. For the evaluation of the ray equation from Fermat's principle, see Chapter 6.
3. E. Hecht and A. Zajac, *Optics*, Addison-Wesley, Reading, Mass., 1974.
4. M. Born and E. Wolf, *Principles of Optics*, Pergamon, New York, 1964, p. 315.
5. A. Yariv, *Quantum Electronics*, Wiley, N.Y., 1967, p. 86.
6. For example, *Melles Griot Optics Guide* 2.
7. J. W. Goodman, *Introduction to Fourier Optics*, McGraw-Hill, New York, 1968, Chap. 3.
8. I. S. Sokolnikoff and R. M. Redheffer, *Mathematics of Physics and Modern Engineering*, McGraw-Hill, New York, 1966.
9. O. Kafri, S. Kimel, and J. Shamir, *J. Quant. Elect.*, **QE-8**, 295 (1972).

SYMBOLS

a	aperture size	$\gamma(\tau)$	degree of coherence
c	speed of light in vacuum	Γ	autocorrelation function

f	focal length	$\delta\nu$	bandwidth
k	wave number	$\delta\Omega$	solid angle
L	cavity length (of laser)	δx	spatial resolution
m	number of longitudinal modes in a pulse	$\Delta\nu$	width of radiation mode
		τ	time gap
n	refractive index		
n	number of transverse modes of radiation		
N	number of radiation modes		
R	spatial resolution		
t	exposure time		
T	temporal extent, duration		
V	volume		

4

Holographic vs. Moire Contouring of Three-Dimensional Diffusive Objects

4.1. INTRODUCTION TO THE METROLOGY OF DIFFUSIVE OBJECTS

Most of the objects we see around us are optically diffusive, i.e., they scatter light incident on them to all directions at random so that part of the scattered radiation enters our field of view and makes us aware of their presence. Diffusive objects may either be partially transparent, in which case the diffusely transmitted light is measured, or opaque, in which the backscattered light diffusely reflected from the object's surface is viewed. In any case, the optical quality of scattered light is too low to be analyzed by means of classical interferometry, in which the wavefronts compared are nearly planar or spherical and deviate only slightly (in the order of magnitude of a wavelength) from an ideal wavefront. In other words, a high degree of spatial coherence [1] is a prerequisite in interferometric measurements.

Nevertheless, the optical study of diffusive objects is far too important for engineers, scientists, and even physicians, to be discarded. Some applications that require such measurements are changes in the physical dimensions of objects subject to stress, analysis of the vibration and inner motion of objects, and the topographic mapping of three-dimensional objects.

The problem of testing diffusive objects by optical means was solved only in the late 1960s by simultaneously applying two newly developed methods, holographic interferometry and shadow moire (also called projection moire). As will be shown, these methods are complementary; they span different regions of sensitivity. Holo-

53

graphic interferometry requires photography and the advanced technology of highly coherent light sources, whereas shadow moire is a low cost solution applicable when lower sensitivity can be tolerated. Therefore, the contemporaneous appearance of the two methods seems rather coincident.

The first topographic mapping of objects by a moire-based method was probably carried out by Mulot [2] in 1925, but the field remained dormant until 1970 when Meadows et al. [3] and Takasaki [4] independently published their work on shadow moire analysis of 3-D objects. This revived the interest in moire topography to an avalanche of some hundred papers [5], both in optical engineering and medicine (where the largest application is the diagnosis of scoliosis [6]).

Holography, or wavefront reconstruction imaging, was invented by Gabor [7] around 1947, but was first applied to the observation of 3-D diffusive objects by Leith and Upatnieks [8] in 1965. That year also saw the birth of holographic interferometry, first reported by Powell and Stetson [9]. This was a breakthrough in interferometry that made it possible to compare wavefronts containing arbitrary phase changes with interferometric precision. Soon the new approach was also adapted by Hildebrand [10] and Varner [11] to the topographic mapping of 3-D diffusive objects.

4.2. HOLOGRAPHY AND HOLOGRAPHIC INTERFEROMETRY

Throughout this book we draw an analogy between moire-based techniques and interferometry. Following that line, we introduce the shadow moire technique as a long wavelength analog to holographic interferometry. For this purpose, a brief introduction of holography and holographic interferometry is given in the following text.

4.2.1. Holography: Introduction

Holography is a method of storing both the amplitude and phase of a wave on photographic film so that the original wavefront can later be reconstructed merely by illumination of the processed film (i.e., the hologram). The recording stage involves the conversion of the phase and amplitude information into intensity, the only kind of information to which the film is sensitive. This is achieved by forming an interference pattern by superposition of the light scattered by the object with a coherent reference beam. When the developed hologram is illuminated with the same type of light as the reference beam, the original object is reconstructed. If the object was a 3-D body, the image retains all its 3-D properties including parallax. If the wavelength and incidence angle of the reconstructing beam are similar to that of the reference beam used for the recording, the object will reappear at the original size and location. The true image can then be optically processed, e.g., for schlieren, deflectometric, or interferometric measurements, exactly as if it were the real object.

4.2.2. Recording Process

The recording process can be treated analytically as follows [12]. We let ψ_1 and ψ_2 designate the reference and the object wave, respectively, at the hologram plane (an xy plane), and ϕ_1 and ϕ_2 designate their phases so that

$$\psi_1(x, y) = \psi_{01} e^{-i\phi_1(x, y)},$$
$$\psi_2(x, y) = \psi_{02}(x, y) e^{-i\phi_2(x, y)}. \tag{4.1}$$

ψ_{01} is a constant amplitude, whereas both phases ϕ_1 and ϕ_2 and the amplitude of the object wave ψ_{02} are a function of the coordinates. The irradiance of the interference pattern $I(x, y)$, obtained by the superposition of ψ_1 and ψ_2 incident on photographic film is

$$I(x, y) = |\psi_1 + \psi_2|^2 = |\psi_{01}|^2 + |\psi_{02}(x, y)|^2 + \psi_1^*\psi_2 + \psi_1\psi_2^*$$
$$= I_1 + I_2(x, y) + 2\psi_{01}\psi_{02}(x, y)\cos[\phi_1(x, y) - \phi_2(x, y)]. \tag{4.2}$$

The transmittance amplitude t_f of the developed hologram consists of a constant bias t_b contributed by the uniform I_1 and a position-dependent term. It is proportional to the product of the exposure time and film sensitivity, given by β', assuming that the film response is linear, namely,

$$t_f(x, y) = t_b + \beta'(I_2(x, y) + \psi_1^*\psi_2 + \psi_1\psi_2^*). \tag{4.3}$$

4.2.3. Reconstruction

In the reconstruction stage the hologram is illuminated by a coherent uniform wave $\psi_3(x, y)$. The light transmitted by the hologram is

$$\psi_3(x, y)t_f(x, y) = t_b\psi_3(x, y) + \beta'I_2(x, y)\psi_3(x, y)$$
$$+ \beta'\psi_1^*\psi_2\psi_3 + \beta'\psi_1\psi_2^*\psi_3$$
$$\equiv U_1 + U_2 + U_3 + U_4. \tag{4.4}$$

If ψ_3 is an exact duplication of the original reference wave ψ_1, the third term becomes

$$U_3 = \beta'|\psi_1|^2\psi_2(x, y) \equiv \beta'I_1\psi_2(x, y), \tag{4.5}$$

namely, U_3 is, up to a constant factor, an exact replica of the original wavefront ψ_2. If, on the other hand, ψ_3 is the complex conjugate of the reference wave ($\psi_3 = \psi_1^*$), the fourth term becomes

$$U_4 = \beta'I_1\psi_2^*(x, y), \tag{4.6}$$

which is proportional to the conjugate of the original object wave. While U_3 is a true image of the object (although in imaging terminology it is regarded as a virtual image), U_4 (the real optical image) gives a pseudoscopic inside-out version of the original object.

Spatial separation between the two twin images U_3 and U_4 is achieved by Leith–Upatnieks (offset-reference) holography in which the reference is introduced at an offset angle (Fig. 4.1a), rather than being colinear with the object–film axis as originally suggested by Gabor. If the offset angle is larger than the divergence angle of the object beam, then the desired image U_3 is fully separated from the other spurious images (Fig. 4.1b).

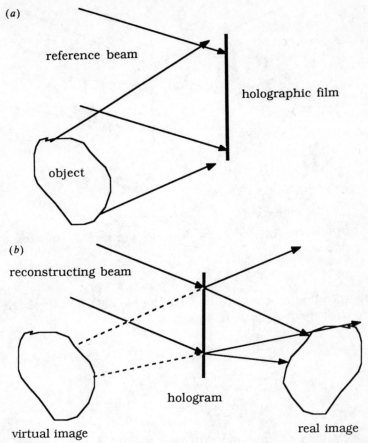

Figure 4.1. Setup for (*a*) recording and (*b*) reconstruction of an offset-reference (Leith–Upatnieks) hologram.

4.2.4. Holographic Interferometry.

In order to obtain difference mapping, the object is returned to its original position relative to the hologram plane for performing the reconstruction (Fig. 4.2). The relocated object's wavefront ψ_4 is given by

$$\psi_4(x, y) = \psi_{04}(x, y)e^{-i\phi_4(x, y)}. \qquad (4.7)$$

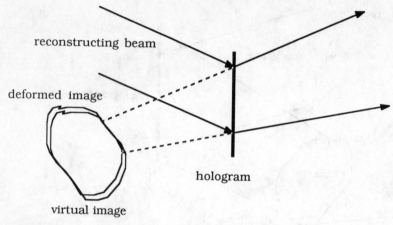

reconstructing beam

deformed image

hologram

virtual image

Figure 4.2. Setup for holographic interferometry (the reconstruction stage), i.e., mapping differences.

By superposition of ψ_4 with the U_3 term, we obtain an interference pattern with the irradiance distribution

$$
\begin{aligned}
I(x, y) = \left\langle (\psi_4 + U_3)^2 \right\rangle &= |\psi_4|^2 + |U_3|^2 + \psi_4^* U_3 + \psi_4 U_3^* \\
&= I_4 + |U_3|^2 + 2\beta' \psi_{04}(x, y) I_1 \psi_{02}(x, y) \\
&\times \cos[\phi_4(x, y) - \phi_2(x, y)].
\end{aligned} \tag{4.8}
$$

Thus, by superimposing the virtual image of the object with the object itself we obtain an interference pattern differing from that produced by a classical interferometer only by the modulation (or contrast) M [12], where

$$
M = \frac{I_{\max} - I_{\min}}{I_{\max} + I_{\min}} = \frac{\beta' I_1 \Psi_{04} \Psi_{02}}{I_4 + \beta'^2 I_1^2 I_2 + \beta' I_1 \Psi_{04} \Psi_{02}}. \tag{4.9}
$$

The low contrast can be adjusted to a maximum by reducing the illumination of the object and increasing that of the reconstruction beam.

This type of measurement is termed *real time* holographic interferometry because it enables the observation of object changes in real time, just as in conventional interferometry. The processed hologram acts as both a beam splitter and a wave front combiner, and compares wave fronts in real time exactly like a classical interferometer.

4.2.5. Double Exposure, Contouring, and Sandwich Holography

Another mode of operation is double exposure holographic interferometry in which the object is recorded twice, either by illumination from different angles or by inducing a certain change in the object between the two exposures, such as applying stress [13]. In the specific example of 3-D object contouring, holographic interferometry supplies a 1 : 1 3-D image of the recorded object with a constant range of contours superimposed on it. The contours are lines of intersection of the object with a set of equidistant planes perpendicular to the line of sight [14].

Contouring is usually done either by illumination of the object with two mutually coherent point sources separated in space [15] or by the use of two coincident plane waves at two frequencies [11]. A modification of the first method is the sandwich-holography contouring suggested by Abramson [16], in which two holograms are made, one with each light source, and a sandwich is made from them. When the sandwich is rotated, the two image waves are mutually sheared and form a set of fringes. The ability to rotate the contouring planes makes it possible to distinguish between hills and valleys.

In the two-frequency method, both the reference beam and the object beam consist of two close wavelengths so that two separate holograms are formed in the same film. When the processed film is illuminated by a single-frequency light, two 3-D images that differ slightly in size and position are formed and interfere with each other. If $\Delta\phi$ is the phase difference between the two wavelengths on passing twice a distance z, namely,

$$\Delta\phi = \frac{4\pi nz}{\lambda_1} - \frac{4\pi nz}{\lambda_2} = \frac{4\pi nz}{\lambda_1\lambda_2}(\lambda_2 - \lambda_1), \qquad (4.10)$$

where n is the refractive index, then the spacing between two fringes (i.e., $\Delta\phi = 2\pi$) is

$$\Delta z = \frac{\lambda_1 \lambda_2}{2n(\lambda_1 - \lambda_2)}. \qquad (4.11)$$

Sometimes we are not interested in the absolute topography of an object but in the difference between two positions of the object. For example, Fig. 4.3 is a holographic interferogram of a lamp with the power on and the reference hologram is the same lamp with the power off.

It is important to notice that the three-dimensionality of the holographic image is insignificant for metrologic purposes. Hologra-

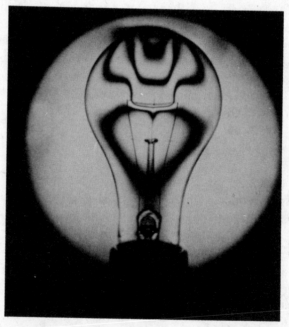

Figure 4.3. A holographic interferogram of differences. The fringes represent optical path differences between a light bulb with the power on and a light bulb with the power off. From G. B. Brandt, Holographic Interferometry, *Handbook of Optical Holography*, H. J. Caulfield, Ed., p. 472.

phy is just a means of producing interference patterns with scattered light that could not be obtained by classical interferometry.

4.3. SHADOW MOIRE AND THE GRATING HOLOGRAM

The major achievement of holography is the ability to store a 3-D picture on a 2-D screen. This is done by creating an interference pattern that converts the parameters of the time-varying electric field (i.e., the phase and the amplitude) into light intensity, to which the photographic film is sensitive. Three-dimensional object contouring by moire methods avoids the complexity of the holographic process since the spatial grating, which in holographic interferometry is provided by the halogram, is merely a distorted linear grid of alternating transparent and opaque stripes, such as the shadow of a Ronchi ruling projected on a 3-D object. Obviously, the density that such a grating can acquire is limited. Usually it does not exceed 50 lines/mm and therefore the sensitivity of a moire-based contouring method is 1 or 2 orders of magnitude lower than that of holographic interferometry.

The basic shadow moire setup comprises a light source (see Fig. 4.4), that projects a shadow of a linear grating tilted at an angle α to the beam direction onto a 3-D diffusive object. A camera or other recording device is placed perpendicular to the grating and a distorted grating is observed. The amount of distortion is proportional to $h(x, y)\tan \alpha$ where $h(x, y)$ is the height variation function.

To determine the fringe position, the grating can be substituted by a line grid, which is described by a set of indicial equations as

$$y = mp, \qquad m = 0, \pm 1, \pm 2 \ldots . \qquad (4.12)$$

If the shadow grating is distorted, i.e., the lines deviate from linearity by an amount $h(x, y)\tan \alpha$, then the equations acquire the form

$$y + h(x, y)\tan \alpha = np, \qquad n = 0, \pm 1, \pm 2, \ldots . \qquad (4.13)$$

Since both gratings, the straight one and its deformed shadow, are parallel (i.e., the intersection angle is $\theta = 0$), then their superposition

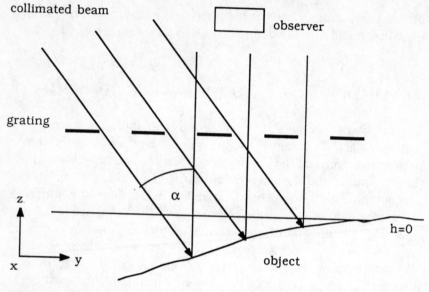

Figure 4.4. Setup for shadow moire. A collimated light beam projects the shadow of a grating onto a diffusive object. The beam is projected at an angle α.

results in an infinite fringe-type moire pattern. If the two overlapping gratings were identical up to a constant phase difference, the infinite fringe would have a gray level varying from 50% transmittance for a phase shift zero to total darkness when the phase shift is π. However, if the gratings are not identical $[h(x, y)\tan \alpha \neq 0]$ as in the case of the 3-D object [Eqs. (4.12) and (4.13)], then by substituting $n - m$ by l we obtain

$$h(x, y) = lp \cot \alpha. \tag{4.14}$$

Equation (4.14) presents the topographic contour map of the object with a height increment $p \cot \alpha$ between successive fringes.

An example of such a contour map is given in Fig. 4.5, which shows a human body photographed through a grating by a camera whose field of view is tilted with respect to the incident beam.

Actually, the 3-D information [i.e., the $h(x, y)$ map] is fully stored on the distorted shadow of the grating and the unperturbed

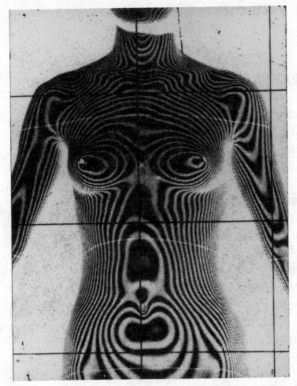

Figure 4.5. Infinite fringe moire pattern of a human body. From H. Takasaki, *Appl. Optics* **9**, 1469 (1970).

grating is only used for obtaining a coarse grain, easy to interpret, moire pattern. In this respect the role of the distorted grating is equivalent to that of the hologram used in holographic interferometry. Takasaki coined the term *grating hologram* for the distorted grating cast on the object [4] to emphasize the analogy. The unperturbed grating used to obtain the contour map is equivalent to the reference beam used to reconstruct a hologram. The grating holograms lack, as a matter of fact, one of the most remarkable properties of a genuine hologram, i.e., the ability to reconstruct a 3-D object upon illumination by a reference beam. However, stereoscopic vision, which can also be achieved by manipulation of 2-D pictures, is generally meaningless in metrology. Thus, the only practical

difference between shadow moire and holographic interferometry is
in the sensitivity.

4.4. DEFERRED ANALYSIS IN SHADOW MOIRE

In a manner similar to holographic interferometry, shadow moire
can also be practiced both as a real time method, as previously
shown, and in various modes of deferred analysis. Delayed analysis
offers a variety of applications that are not possible in real time. For
delayed analysis the grating hologram alone is recorded, illuminated
by a collimated beam, point source, or simply a slide projector (see
Fig. 2.4a). The use of point source (or slide projector) illumination
introduces imperfections into the contour map, but on the other
hand enables the recording of large objects, for example, the grating
hologram of a house (Fig. 4.6). If the same source is used to produce
the reference grating, the noise can be easily removed.

Figure 4.7a shows the setup for recording the grating hologram of
an object. A collimated beam projects a Ronchi ruling on a 3-D
diffusive object and the grating hologram is described by

$$y + h_1(x, y)\tan \alpha = np. \tag{4.15}$$

A transparency of the developed film is then positioned in place of

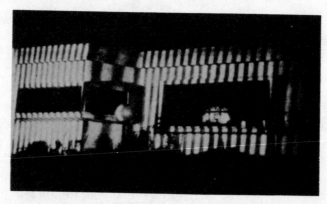

Figure 4.6. Grating hologram of a house.

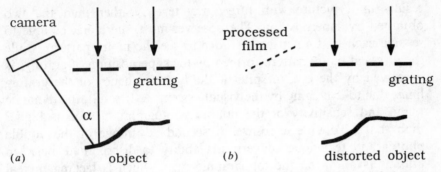

Figure 4.7. Setup for (a) recording and (b) reconstruction of a grating hologram.

the film plane (Fig. 4.7b). Let us assume that the object was subject to a deformation after the first exposure and is now described by

$$y + h_2(x, y)\tan \alpha = mp. \qquad (4.16)$$

The contour map observed through the transparency is a map of differences, namely,

$$h_2(x, y) - h_1(x, y) = lp \cot \alpha. \qquad (4.17)$$

This mode of operation is analogous to real time holographic interferometry. It is extremely useful in following changes of an object due to stress applied to it or to inner motion.

Another way of obtaining a contour map of differences is by double exposure of the same film, where the object suffers a certain deformation between the two exposures.

A third version is in analogy to the sandwich-holography contouring discussed in Sec. 4.2.5. Two separate photographs of the object are taken, one at a reference position and the other after some change is introduced. Then the developed transparencies of the two grating holograms are superimposed. This mode has several advantages over the double exposure technique. First, the superposition operation is represented mathematically by a multiplication of two transmittance functions, as does superposition of the gratings themselves (see Chapter 2 for more details). Double exposure, on the other hand, is equivalent to the addition of light intensities and results in a picture with three gray levels, rather than the two obtained by superposition. The observer must, therefore, be able to resolve each grating stripe in order to see the moire pattern, while the classical moire pattern formed by the superposition of gratings is perceived by the eye, in spite of the high frequency of the grating lines, due to averaging by the visual system. As a result, the dynamic range and sensitivity of the double photograph method is higher than that of double exposure. A second advantage of the double photograph technique concerns its ability to discriminate between hills and valleys on the topographic map, which is lacking in real time or double exposure methods shown before.

4.5. SLOPE DIRECTION DETERMINATION IN SHADOW MOIRE

4.5.1. Infinite Fringe Shadow Moire

The contour map obtained by shadow moire is a convenient tool for quick determination of the hilliness of a surface. However, the information provided by an infinite fringe moire pattern is rather incomplete. It does not give the incrementation sign, but only its absolute value. As a result, not only can we not distinguish between hills and valleys, we even cannot tell whether a sequence of contours represents a monotonous change in height, or rather an alternating height change (a sawtooth shaped or wavy surface). The problem of discriminating between hills and valleys can be solved by producing a finite fringe moire pattern instead of the infinite fringe mode of the shadow moire technique.

4.5.2. Finite Fringe Shadow Moire

In the finite fringe mode of operation [17], the two gratings, the deformed one and the reference grating, are mutually inclined during the superposition, forming a small intersection angle θ between the directions of the lines. If the two gratings are identical, a pattern of straight fringes perpendicular to the bisector of θ is formed. Assume that the reference grating and the distorted grating lying in the xy

plane are symmetrically inclined about the x axis at angles $\theta/2$ and $-\theta/2$, respectively. Since the camera is located at angle α to the z axis, the images of the reference grating and the distorted shadow formed in the film plane are given by the equations

$$y \cos \theta/2 = x \sin \theta/2 + mp,$$
$$[y + h(x, y)\tan \alpha]\cos \theta/2 = -x \sin \theta/2 + np, \qquad (4.18)$$

respectively. If the two gratings were identical [i.e., $h(x, y) = 0$], the solution would be an array of equally spaced straight fringes progressing in the x direction with pitch $p' = p/(2 \sin \theta/2) \approx p/\theta$. When one of the gratings is distorted as in Eq. (4.18), the fringes shift from their unperturbed position by an amount x', where

$$x' = \frac{h \tan \alpha \cot \theta/2}{2} \approx \frac{h(x, y)\tan \alpha}{\theta}, \qquad (4.19)$$

namely, the fringe shift is linearly proportional to the difference in height h, and a negative slope results in an inverted fringe shift direction compared to a positive slope. An example of a finite fringe mapping of an object is given in Fig. 2.4c. Figure 2.4b shows the infinite fringe moire pattern, i.e., a contour map, of the same object. Finite fringe patterns can be obtained either by the double exposure technique, in which the reference grating is slightly rotated in the xy plane between the two exposures, or by double photography, in which the original grating does not need to be touched during the recording stage. Instead, one can freely rotate the two transparencies of the processed film to obtain the desired results, as in sandwich holography.

4.6. NOISE REDUCTION IN MOIRE CONTOURING BY DOUBLE EXPOSURE

In addition to the advantages already discussed of double exposure techniques over the basic shadow moire approach, there is also the noise reduction property of double exposure. In holographic interfer-

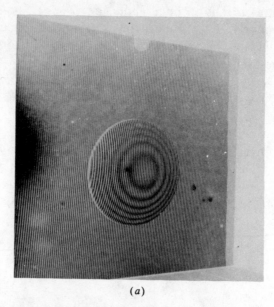

(a)

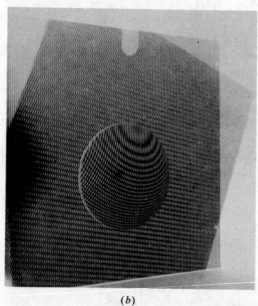

(b)

Figure 4.8. (a) Contour map of a diffusive dome. (b) Finite fringe mapping of the same dome. The fringe distortion is proportional to the height at the location of the fringe.

ometry the problem of image quality degradation by aberrating media present along the optical path can be solved in several ways as discussed by Goodman [ref. 12, pp. 261–268). In one of those techniques the object is recorded on the hologram through the inherent aberrating medium, and that medium is also present at the same distance from the hologram during the reconstruction. The net effect is total cancellation of the disturbance.

In moire contouring we adopt the same idea, i.e., the aberrating medium (for example, a transparent case covering the object) is present at the same position in both exposures, and therefore its detrimental effect is eliminated. This can be easily explained if the two gratings of Eq. (4.18) suffer from the same distortion, namely,

$$[y + h(x, y)\tan\alpha]\cos\theta/2 = x\sin\theta/2 + mp,$$
$$[y + h(x, y)\tan\alpha]\cos\theta/2 = -x\sin\theta/2 + np. \quad (4.20)$$

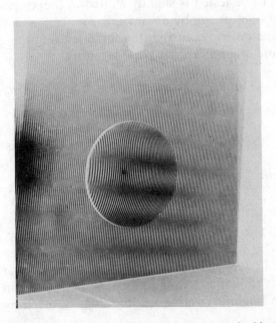

Figure 4.9. Straight moire fringes produces by finite fringe double exposure of the dome of Fig. 4.8.

The solution is again a set of straight fringes with spacing $p' \cong p/\theta$. In the infinite fringe mode the result is a uniform gray level throughout the overlapping area. The effect is demonstrated in Figure 4.8. Figure 4.8a is an infinite fringe mapping of a dome and Fig. 4.8b is a finite fringe double exposure photograph of the same dome. Figure 4.9 shows what happens when two identical grating holograms of the same dome are superimposed with a small angle between them. As predicted by Eq. (4.20), straight fringes are obtained.

If the two grating holograms do not overlap completely, another phenomenon is observed. A slight mutual shearing of the gratings results in a map of lateral derivatives of the height, i.e., a slope map of the object, as will be discussed in the next section.

4.7. SLOPE CONTOURING BY SHEARING SHADOW MOIRE TECHNIQUES

If the object to be tested is slightly shifted in the xy plane between two exposures, let us say by an amount δy, then, assuming that no other deformation has occurred, the finite fringe grating holograms of the first and second exposures are given by

$$[y + h(x, y)\tan \alpha]\cos \theta/2 = x \sin \theta/2 + np,$$
$$[y + \delta y + h(x, y + \delta y)\tan \alpha]\cos \theta/2 = -x \sin \theta/2 + mp, \quad (4.21)$$

respectively. The solution, already slightly rearranged, is

$$\frac{\partial h}{\partial y} = \frac{h(x, y + \delta y) - h(x, y)}{\delta y} \approx \frac{x'\theta}{\delta y \tan \alpha}, \quad (4.22)$$

where x' is the fringe shift from its unperturbed position (see Chapter 2). This states that the fringe shift x' is linearly proportional to the slope in the y direction. Shearing of parallel grating holograms, i.e., in the infinite fringe mode, was discussed in ref. 18. In this case, the height derivative, i.e., the slope, is given by

$$\frac{\partial h(x, y)}{\partial y} = \frac{h(x, y + \delta y) - h(x, y)}{\delta y} \approx \frac{lp}{\delta y \tan \alpha}. \quad (4.23)$$

Figure 4.10.   Finite fringe slope mapping of a diffusive dome.

The increment between two equiderivative fringes is given by

$$\left[\frac{\partial h(x, y)}{\partial y}\right]_{incr} = \pm \frac{p}{\delta y \tan \alpha} \tag{4.24}$$

in which the plus/minus sign reflects the ambiguity in the slope direction when working with infinite fringe patterns. Figure 4.10 shows a lateral derivative finite fringe map of a dome. The holographic analog to shearing moire is lateral shearing interferometry [19], mentioned on several occasions in this book. Notice that shearing moire is a recursive method. By sheared superposition of the contour map of slopes with itself one can obtain the second derivative of the height, i.e., the map of curvatures of the object (as was discussed in Chapter 2).

4.8. VIBRATION ANALYSIS

Vibration analysis is an important application of shadow moire that enables the monitoring of an object's vibration in real time. The analysis can be done in two ways. The first is a real time method, that records the shadow moire pattern of a vibrating object with a detector. The light intensity of the moire fringes is, within one fringe period, linearly proportional to the fringe shift and, therefore, to the object's motion. As a result, a pointwise real time description of the motion is obtained [20]. Such an analysis is more complicated in holography because of the sinusoidal shape of the fringe profile.

The other approach common in holography [21], speckles interferometry [22], and shadow moire [23, 24] is to record the entire pattern of the object. In the areas where the vibration of the object has nodes, the fringes appear sharp. In the regions of high vibrations, they appear blurred. A long exposure photograph produces a time averaged vibration pattern of the object. Holography and speckle interferometry techniques are suitable for analysis of vibrations with amplitude in the wavelength regime, while shadow moire can be applied to higher amplitude vibrations from 0.01 mm and above.

4.9. SUMMARY

The topographic mapping of 3-D objects by shadow moire and a variety of other methods derived from it has been discussed. Throughout the presentation we have drawn an analogy between moire and the holographic contouring methods. To summarize, when the sensitivity requirements are moderate, i.e., when the height difference resolution required does not exceed 0.01 mm, moire-based methods provide a low cost, easily applicable solution to the contouring problem. If better resolution (of the order of magnitude approaching one wavelength, i.e., $\sim 0.5 \ \mu$m) is required, holographic interferometry is the preferred solution. This solution is costly since a highly coherent light source (singe-mode laser) of sufficient intensity, high quality optics, and a mechanically stabilized optical bench or very short pulses, are indispensible prerequisites for holographic experiments. A reader interested in a comprehensive review of the techniques used in shadow moire is referred to ref. 5.

REFERENCES

1. For a detailed discussion on the coherence of light see Chapter 3.

2. M. Mulot, Application du Moire a l'etude des Deformations du Mica, *Revue Opt. Theor. Instrum.* **4**, 252 (1925).

3. D. M. Meadows, W. O. Johnson, and J. B. Allen, Generation of Surface Contours by Moire Patterns, *Appl. Opt.* **9**, 942 (1970).

4. H. Takasaki, Moire Topography, *Appl. Opt.* **9**, 1469 (1970).

5. J. R. Pekelsky and H. C. Van Wijk, in *Handbook of Non-Topographic Photogrammetry*, H. M. Karara, ed., American Society of Photogrammetry and Remote Sensing, Chap. 15.

6. B. Drerup, W. Frobin, and E. Hierholzer, Eds., *Moire Fringe Topography and Spinal Deformity*, Gustav Fischer Verlag, Stuttgart, 1983.

7. D. Gabor, A New Microscopic Principle, *Nature* **161**, 777 (1948).

8. E. N. Leith and J. Upatnieks, Wavefront Reconstruction with Diffused Illumination and Three-Dimensional Objects, *J. Opt. Soc. Am.* **54**, 1295 (1964).

9. R. L. Powell and K. A. Stetson, Interferometric Vibration Analysis of Three-Dimensional Objects by Wavefront Reconstruction, *J. Opt. Soc. Am.* **55**, 612 (1965).

10. B. P. Hildebrand, General Analysis of Contour Holography, Ph.D. Dissertation, University of Michigan, Ann Arbor, 1967.

11. J. R. Varner, Multiple Frequency Holographic Contouring, Ph.D. Dissertation, University of Michigan, Ann Arbor, 1971.

12. J. W. Goodman, *Introduction to Fourier Optics*, McGraw-Hill, N.Y., 1968, Chap. 8.

13. G-B. Brandt, Holographic Interferometry, in *Handbook of Optical Holography*, H. J. Caulfield, Ed., Academic, N.Y., 1979, Chap. 10.4.

14. J. R. Varner, Holographic Contouring Methods, in *Handbook of Optical Holography*, H. J. Caulfield, Ed., Academic, N.Y., 1979, Chap. 10.10.

15. B. P. Hildebrand and K. A. Haines, Multiple Wavelength and Multiple-Source Holography Applied to Contour Generation, *J. Opt. Soc. Am.* **57**, 155 (1967).

16. N. Abramson, *Appl. Opt.* **15**, 200 (1976).

17. A. Livnat and O. Kafri, Finite Fringe Shadow Moire: Slope Mapping of Diffusive Objects, *Appl. Opt.* **22**, 3232 (1983).

18. O. Kafri, A. Livnat, and E. Keren, Optical Second Differentiation by Shearing Moire Deflectometry, *Appl. Opt.* **22**, 650 (1983).

19. Y. Y. Hung, J. L. Turner, M. Tafralian, J. Der Hovanessian, and C. E. Taylor, *Appl. Opt.* **17**, 128 (1978).
20. O. Kafri, Y. B. Band, T. Chin, D. F. Heller, and J. C. Walling, Real Time Moire Analysis of Diffusive Objects, *Appl. Opt.* **24**, 240 (1985).
21. P. Shajenko, Holographic Testing of Loud Speakers, *J. Acoust. Soc. Am.* **53**, 1061 (1973).
22. F. P. Chiang and R. M. Jung, Vibration Analysis of Plate and Shell by Laser Speckle Interferometry, *Opt. Acta* **23**, 977 (1976).
23. J. Der Hovanessian and Y. Y. Hung, Moire Contour-Sum Contour-Difference, and Vibration Analysis of Arbitrary Objects, *Appl. Opt.* **10**, 2734 (1971).
24. R. Ritter and H. J. Meyer, Vibration Analysis of Plates by Time Averaged Projection Moire Methods, *Appl. Opt.* **19**, 1630 (1980).

SYMBOLS

h	height difference	Δx	spacing between two fringes
I	irradiance	α	viewing angle
m, n, l	indicial equation constants	β'	(exposure time) × (film sensitivity)
M	modulation	λ	wavelength
n	refractive index	θ	grating's angle
p	grating pitch	ψ	wave
t_f	transmittance amplitude		
t_b	bias term of transmittance amplitude		
U	holographic image		
x'	fringe shift		

5

Moire Analysis of Strain

In Chapter 4 we presented several methods for geometrical analysis of 3-D diffusive objects. From these methods we can obtain a contour map of the shape function $f(x, y, z) = 0$, maps of geometrical differences between two objects, or the changes within an object in time. The most impressive aspect of the optical methods, i.e., moire technique and holography, is that no physical contact with the object is required. However, these techniques do not provide information about all the changes that might occur in a solid object in time. Suppose that we have a 2-D flat surface $f(z = 0)$. If a stress is applied in the xy plane, distortions, representing strain, will appear. These distortions will not be mapped by either holography or shadow moire. Nevertheless, a sizable group of scientists and mechanical engineers are interested in measuring these distortions and their relation to stress. They are responsible for the development of many techniques for strain analysis based on the moire effect described in the books written by Durelli and Parks [1] and Theocaris [2]. Recently Post [3, 4] introduced the interferometric methods of in-plane topographic analysis for measuring strain. In this chapter we will first discuss the classical moire techniques and moire interferometry. This will be followed by a discussion on applying moire deflectometry to direct measurement of strain.

5.1. BACKGROUND ON STRESS AND STRAIN

Before we discuss the principle of measuring strain by the moire effect, we will briefly review the concept of strain. An elaborate discussion on the subject is given in the books written by Durelli and Parks [1] and Theocaris [2]. Here we will discuss only the basics.

75

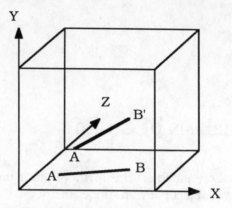

Figure 5.1. A cube subject to stress.

The cube of Fig. 5.1 can change its volume due to forces applied in many directions. The line AB may be transformed to a line $A'B'$, or may even break into several pieces. The line $A'B'$ may differ in both location direction and length. The tensor of strain is designated as ϵ_{ij}. The diagonal terms of the strain tensor ϵ_{xx}, ϵ_{yy}, and ϵ_{zz} are called the normal strain and are defined as

$$\epsilon_{xx} = \partial u/\partial x,$$
$$\epsilon_{yy} = \partial v/\partial y,$$
$$\epsilon_{zz} = \partial w/\partial z, \tag{5.1}$$

where u, v, and w are the components of the displacement vector in x, y, and z directions, respectively. Thus, the strain is defined as the derivative of the displacement, and therefore we have a 3×3 tensor. As can be understood intuitively, u may be changed either by applying a force or as a result of stress in the x, y, or z direction. A shear-type strain is defined as the change in the right angle in a hypothetical cube due to deformation. The shear strain is defined as the sum of the two mixed terms, for example,

$$\epsilon_{xy} = (\partial v/\partial x + \partial u/\partial y)/2. \tag{5.2}$$

If all the tensor elements were independent, a situation could occur where not only will deformation be present, but the material will no longer be continuous. Since strain analysis is defined mostly for isotropic materials, the continuity differential equations result in symmetry rules that reduce the number of independent elements,

$$\epsilon_{xy} = \epsilon_{yx}, \qquad \epsilon_{yz} = \epsilon_{zy}, \qquad \epsilon_{xz} = \epsilon_{zx}.$$

This leaves us with three terms for normal strain and three terms for shear strain:

$$\epsilon_{ij} = \begin{bmatrix} \epsilon_{xx} & \epsilon_{xy} & \epsilon_{xz} \\ \epsilon_{xy} & \epsilon_{yy} & \epsilon_{yz} \\ \epsilon_{xz} & \epsilon_{yz} & \epsilon_{zz} \end{bmatrix}. \tag{5.3}$$

This tensor may be diagonalized to the tensor

$$\epsilon'_{ij} = \begin{bmatrix} \epsilon_{11} & 0 & 0 \\ 0 & \epsilon_{22} & 0 \\ 0 & 0 & \epsilon_{33} \end{bmatrix}. \tag{5.4}$$

These new directions 1, 2, and 3 are called the principal axes of the material and ϵ_{11}, ϵ_{22} and ϵ_{33} are called the principal strain components.

Strain is a result of stress and is proportional to it. A liquid without shear can experience only three stress components Xx, Yy, and Zz. In solids one also can apply shear stress Xy, Xz, and Yz, namely, by applying a force in the x direction, a tension in the y direction is formed. The stress is also a 3×3 tensor with similar properties to the strain tensor. The relationship between stress and strain is a property of the material. In principle, lacking symmetry to relate between two 3×3 tensors, a tensor of $3 \times 3 \times 3 \times 3$, which has 81 terms, is required. However, from symmetry considerations, only 15 independent terms can be measured.

5.2. STRAIN MEASUREMENTS

Since in-plane strain measurements require a reference taken prior to the time of deformation, all strain analyses should be done by "touching" the measured object. The most common device for strain analysis is the strain gauge. This is a tiny resistor glued to the surface that changes its resistance with a change in its dimensions. An electrical device translates the change in current to a change in dimensions. The popularity of the strain gauge is due to its simplicity and its ability to measure both in-plane and out-of-plane strain. However, its serious drawback is the lack of mapping.

Here we will show the moire approach of mapping strain, which is, in fact, very simple. A grating is glued or etched on a flat test surface and by applying a stress to the surface, the grating will be distorted. In general, if the surface $f(x, y)$ is transformed to a surface $f(x + \delta x, y + \delta y)$, then the grating which is glued onto the surface

$$y = np, \qquad n = 0, \pm 1, \pm 2, \ldots, \tag{5.5}$$

will be transformed to

$$y + v(x, y) = mp, \qquad m = 0, \pm 1, \pm 2, \ldots, \tag{5.6}$$

where $v(x, y)$ is the displacement of the point (x, y) in the y direction.

The infinite fringe moire pattern obtained by superposition of the grating described by Eq. (5.6) with an undistorted grating of Eq. (5.5) is the solution of the two sets of equations, i.e.,

$$v(x, y) = lp, \qquad l = m - n = 0, \pm 1, \pm 2, \ldots . \tag{5.7}$$

This is a contour map of the distortions in the y direction. To obtain the contour map in the x direction, the grating (5.5) must be printed perpendicularly and the corresponding equations result in

$$u(x, y) = lp. \tag{5.8}$$

It is now possible to find the derivatives from the slopes of the displacements contours or by shearing moire technique (see Chapter 2).

The shearing is done by superimposing the contour map of the displacements with a shifted replica of itself. Suppose that we have the contour map

$$v(x, y) = np. \tag{5.9}$$

with its shifted replica, i.e.,

$$v(x, y + \delta y) = mp. \tag{5.10}$$

The fringe pattern will be,

$$v(x, y + \delta y) - v(x, y) = lp. \tag{5.11}$$

By dividing both sides of (3.11) by δy we obtain

$$[v(x, y + \delta y) - v(x, y)]/\delta y \cong \partial v/\partial y = lp/\delta y, \tag{5.12}$$

namely, a contour map of the partial derivatives incremented by $p/\delta y$. This technique is not practical in most cases because only a few fringes exist in the original contour map and the moire pattern, which is a moire of moire, is not smooth and accurate. By using a finite fringe map of the distortion, the amount of fringes is increased without losing any information.

Suppose that we take the distorted grating (5.6) and superimpose it with a reference grating (5.5) at a small angle θ. We rewrite the equations of the two gratings as

$$y \cos(\theta/2) = -x \sin(\theta/2) + np \tag{5.13}$$

and

$$[y + v(x, y)]\cos(\theta/2) = x \sin(\theta/2) + mp. \tag{5.14}$$

The displacements are obtained as a finite fringe map (see Chapter 2), and Eqs. (5.13) and (5.14) yield

$$v(x, y)\cos(\theta/2) = 2x \sin(\theta/2) + lp. \tag{5.15}$$

For small θ, the cosine term may be assumed to be unity and the sine term may be replaced by the angle itself. Therefore, Eq. (5.15) reduces to

$$v(x, y) \cong x\theta + lp. \tag{5.16}$$

Equation (5.16) can be rewritten as

$$x - \frac{v(x, y)}{\theta} \approx lp',$$
(5.17)

which is a new grating similar to that of Eq. (5.6). Its pitch p' is magnified by a factor $1/\theta$ and its distortion is proportional to $v(x, y)/\theta$. In addition, the stripes of the grating are perpendicular to the original grating of (5.6). The pitch can be tuned by simply changing θ, and this new grating will be the base grating for obtaining the *moire of moire* map of the strains. Shifting the grating of (5.17) by a small amount δy results in

$$x - \frac{v(x, y + \delta y)}{\theta} \cong l'p',$$
(5.18)

and by substracting Eqs. (5.18) from (5.17) we obtain

$$v(x, y + \delta y) - v(x, y) = np, \quad n = l' - l.$$
(5.19)

Dividing both sides by δy we obtain a contour map of the normal strain component in the y direction:

$$\epsilon_{yy} = \frac{v(x, y + \delta y) - v(x, y)}{\delta y} = \frac{np}{\delta y}.$$
(5.20)

Equation (5.20) is identical to Eq. (5.12) except that here we have a way to control the number of fringes that are used to obtain the map. A map of ϵ_{xx} can be similarly obtained by rotating the printed gratings by 90°.

5.3. MAPPING SHEAR STRAIN

As we saw in Eq. (5.2), shear strain is the sum of two mixed terms, and therefore we must first map the partial derivatives $\partial v/\partial x$ and $\partial u/\partial y$. The two maps are then combined to give us a map of ϵ_{xy}. The mixed derivatives are similarly obtained by combining shifted patterns. For example, if the finite fringe grating of Eq. (5.17) is combined with its replica displaced by an amount δx, instead of Eq. (5.18), we obtain,

$$x + \delta x - \frac{v(x + \delta x, y)}{\theta} \cong l'p'. \qquad (5.21)$$

Superpositioning the two gratings yields

$$v(x + \delta x, y) - v(x, y) - \theta \, \delta x = np. \qquad (5.22)$$

By dividing both sides by δx we find that

$$\frac{\partial v}{\partial x} \cong \frac{v(x + \delta x, y) - v(x, y)}{\delta x} = \frac{np}{\delta x} + \theta. \qquad (5.23)$$

As before, the fringe increment is p divided by the shift and θ is a constant phase term that can be neglected because it is not a function of x or y. The $\partial u/\partial y$ term can be similarly mapped.

5.4. ANALYSIS OF STRAIN BY MOIRE INTERFEROMETRY

The method of moire analysis is simple and straightforward, however, its sensitivity is limited not only by the pitch of the grating fixed to the strained material, but mainly by the inability to obtain moire fringes of reasonable contrast with two very dense gratings (in which the pitch is equal or smaller than a wavelength). For example, if $p = \lambda$, the moire pattern is totally blurred when the two gratings are displaced by a fraction of a micron, due to diffraction.

In order to overcome this problem Post [3] and Weissman and Post [4] developed a method called moire interferometry that employs a new type of interferometer for measuring in-plane distortions. In Fig. 5.2 the schematics of moire interferometry are described. A diffraction grating is fixed onto a surface under test. If a collimated beam is projected perpendicular to the grating it will be diffracted into several orders at angles $n\lambda/p$ where n is the diffraction order and p is the pitch of the fixed grating. The energy

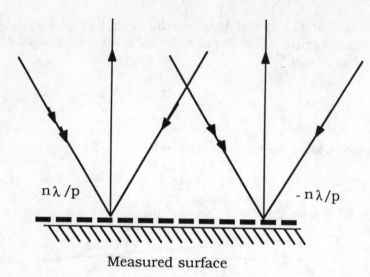

$n\lambda/p$ $-n\lambda/p$

Measured surface

Figure 5.2. The moire interferometer. A grating pitch p is etched onto a surface that is subjected to stress. Two coherent beams are projected onto the grating at angles $\pm n\lambda/p$. They are reflected normal to the grating and interfere to form a fringe pattern related to the strain.

distribution between the various orders is determined by the harmonic structure of the grating reflectance. For example, in Ronchi rulings about 80% of the energy is in the zero and two first orders ± 1. Many commercial gratings are blazed so that most of their energy will be reflected at a given single diffraction order.

In moire interferometry, two mutually coherent beams are projected onto the grating at angles $\phi = \pm n\lambda/p$ so that a large portion of their energy is reflected back perpendicular to the surface. It is interesting to note that the information obtained by the reflected beam, i.e., $+n\lambda/p$ or $-n\lambda/p$, is a wave front distortion and no grating structure appears. If the surface is stretched due to strain, these two beams interfere and the fringe pattern is related to the strain, because the grating is strained along with the object. The diffraction angle is proportional to the pitch of the grating by

$$\phi = n\lambda/p. \tag{5.24}$$

If the pitch is changed locally, part of the beam is deflected by an angle $\delta\phi$ given by

$$\delta\phi = -\left(n\lambda/p^2\right)\delta p, \tag{5.25}$$

where both ϕ and p are functions of x and y.

One can identify $\delta p/p$ as ϵ_{xx}, the normal strain in the previous section. The ray deflection results in phase retardation simply because the wave vectors differ by $\sin(\delta\phi)$ compared to the unperturbed one. In the paraxial approximation $\sin(\delta\phi)$ can be approximated by $\delta\phi$ and the phase retardation is given by $\delta\phi$ where

$$\delta\phi = \lambda/\delta x. \tag{5.26}$$

Since in moire interferometry the two beams are projected in opposite directions, the difference in the wave vector is $2\,\delta\phi$ and the fringes will appear at distances of

$$\delta x = \lambda/2\,\delta\phi. \tag{5.27}$$

By substituting the expression for $\delta\phi$ of (5.25) into Eq. (5.27) we obtain

$$\epsilon_{xx} = \frac{p}{2n\,\delta x}, \tag{5.28}$$

namely, the strain ϵ_{xx} is inversely proportional to the distance between two adjacent fringes or the gradient of the interference pattern.

It is important to understand that, as in conventional moire methods, moire interferometry maps displacements and not the strain. This is because $\epsilon_{xx} = u(x, y)/\delta x$ and the result of the interference of the two beams is exactly as in the moire method, namely, a contour map incremented by the pitch of the grating. The factor $2n$ appears from the use of two opposite beams of the nth diffraction order. As in moire methods, both the strain terms can be obtained by shearing techniques. (For more details, see Patorski et al. [5].)

The maximum sensitivity (or minimum detectable strain) is achieved when the distance between two fringes is equal to the beam aperture a. The minimum displacement detected by moire interferometry is therefore

$$\epsilon_{xx}(\min) = \frac{p}{2na}. \tag{5.29}$$

This method, which applies two beams, eliminates fringes due to out-of-plane distortions. This is because an out-of-plane distortion (slope) on the surface will cause the two beams to deflect in unison and an in-plane distortion will change the pitch of the etched grating and cause the two beams to interfere.

It is seen that no moire effect is really involved in this technique. The name comes from its history, because similar measurements with lower sensitivities applied moire fringes and the governing equations of the two methods are the same. This technique extends the sensitivity range of the moire technique to the diffraction limit.

5.5. ANALYSIS OF STRAIN BY MOIRE DEFLECTOMETRY

Moire deflectometry is a moire-based technique for mapping of ray deflections. Since the change in grating's pitch due to strain causes ray deflections, moire deflectometry addresses itself as the natural tool to directly measure the strain in the diffraction limit. We devote a great part of this book to moire deflectometry and its applications, but for methodical reasons these chapters come later. The reader is advised to read Chapter 6 before continuing this section.

The deflectometric setup for strain measurement is described in Fig. 5.3. As in moire interferometry, a grating is attached to the measured surface and a collimated beam is projected at an angle $\phi = n\lambda/p$. The difference is that only one beam is required and the direction of the beam does not need to be as precise as in the moire interferometric setup. The reflected beam projects the shadow of the grating G_1 onto the grating G_2 and moire fringes are formed. If the gratings G_1 and G_2 are parallel to each other, an infinite fringe pattern is formed. Strains will form a contour map incremented by

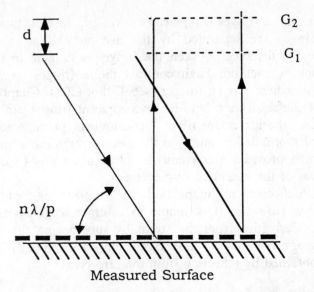

Figure 5.3. The moire deflectometer setup for detecting strain. A grating is etched onto a surface as in Fig. 5.2, but only one beam is projected onto the grating. G_1 and G_2 are the two gratings of the deflectometer and d is their separation. The fringe pattern is a contour map of strains.

the ray deflection p_1/d, where p_1 is the pitch of the gratings G_1 and G_2 and d is the distance between them. Mapping the strain of Eq. (5.25) with increment p_1/d yields

$$\frac{qp_1}{d} = -\frac{n\lambda}{p}\epsilon_{xx} \tag{5.30}$$

or

$$\epsilon_{xx} = \frac{p}{n}\left(\frac{qp_1}{\lambda d}\right), \tag{5.31}$$

where q is the fringe number. By rotating the two gratings of the deflectometer by $90°$, the mixed term $\partial u/\partial y$ can be mapped.

The similarity to moire interferometry is obvious; however, in moire deflectometry, the strain is mapped directly and no additional shearing techniques are required. In addition, the distance between two interference fringes is replaced by $\lambda d/qp_1$. In fact, two adjacent moire fringes are separated by the distance $\lambda d/n$ ($q = 1$) and therefore the distance between the fringes is tunable in the moire deflectometric method. High contrast moire fringes appear when $d = lp_1^2/\lambda$, where l is an integer (see Talbot effect, Chapter 6) and therefore the separation between two adjacent fringes can be tuned by lp_1, i.e., the increment of the deflectometer pitch. Another nice feature of moire deflectometry is the ease in obtaining a finite fringe map, which provides a convenient and accurate way to obtain the derivatives of the measured displacement.

Moire deflectometry maps both the in-plane and out-of-plane distortions. However, it is simple to subtract the unwanted topographical derivative from the strain by superposing the combined deflectogram with that of the surface itself (see Chapter 2). The latter can be obtained by reflection from the zero order.

REFERENCES

1. A. J. Durelli and V. J. Parks, *Moire Analysis of Strain*, Prentice Hall, Englewood Cliffs, N.J., 1970.
2. P. S. Theocaris, *Moire Fringe in Stress Analysis*, Pergamon, London, 1969.
3. D. Post, Developments in Moire Interferometry, *Opt. Eng.* **21**, 458 (1982).
4. E. M. W. Weissman and D. Post, Moire Interferometry Near Theoretical Limit, *Appl. Opt.* **21** 1621 (1982).
5. K. Patorski, D. Post, R. Gzarmek, and Y. Guo, Real Time Optical Diffraction for Moire Interferometry, *Appl. Opt.* **26**, 1977 (1987).

SYMBOLS

a	aperture size
d	gratings' separation, deflectometry
G	grating, deflectometry
n, m, l	indicial equation constants

n	diffraction order
p	pitch of etched grating
p'	pitch of moire pattern
p_1	pitch of deflectometry gratings
q	fringe number
u, v, w	unit vector components in x, y, z directions
ϵ	strain
θ	angle between gratings
λ	wavelength
ϕ	diffraction angle

6

Moire Deflectometry

6.1. INTRODUCTION

In the preceding chapters we demonstrated the classical uses of the moire effect in the metrology of diffusive objects, namely, shadow moire and strain analysis. In shadow moire, the shadow of a grating projected on a diffusive surface interferes with the original grating and the moire pattern formed is a contour map of elevations. In strain analysis, a grating is etched onto an object to be subject to stress. The moire fringes obtained by superposition of the grating, deformed after applying stress, with the original undistorted grating are *isothetics*, i.e., contours of equal displacement. In both applications, the distorted grating (or the shadow) is brought as close as possible to physical contact with the reference grating.

In this chapter we introduce a field of moire effect applications based on a completely different approach. The two gratings are placed apart from each other and the deformation of the moire pattern is caused by ray deflections. When the gratings' stripes are initially parallel, the resulting fringes are contours of equal deflection angle.

In this technique, entitled *moire deflectometry* [1], the object to be tested (either a phase object or a specular surface) is mounted in the course of a collimated beam, followed by a pair of transmission gratings placed at a distance from each other. The resulting fringe pattern, i.e., the *moire deflectogram*, is a map of ray deflections corresponding to the optical properties of the inspected object. Hence, moire deflectometry is a powerful tool for nondestructive optical testing, an area that has been dominated by interferometric techniques. The basis of all interferometric methods is the interference pattern formed by superposition of two or more light waves originating from the same coherent source, but propagating in different paths. The fringe pattern indicates local phase shifts arising from the difference in the optical path traversed by the interfering beams, and the sensitivity to the path difference is a fraction of the wavelength λ (in the visible spectral region, $0.4 < \lambda < 0.7\,\mu m$). This extreme sensitivity, which calls for high mechanical stability, restricts interferometric measurements to low-noise environments and requires high quality optical components and vibration isolated optical benches, even when the actual accuracy demands are not very high.

The treatment of interferometric data is by the mathematically elaborate wave theory. Alternatively, moire-related phenomena can be explained by pure geometrical optics and are treated within the ray tracing formalism. We will show how moire deflectograms are interpreted by the ray tracing approach. Diffraction effects should also be considered, but only as noise, which reduces the fringe contrast and spatial resolution.

The data acquired by moire deflectometry, i.e., the ray deflection map, can be transformed to phase retardation information (as in interferometric measurements) by means of integration. However, as will be shown in Section 7.4, the 3-D analysis of phase objects requires the transverse gradients of the refractive index, rather than the refractive index itself. This is also true in strain analysis where the strains, i.e., the gradients of the deformation, are required. For these applications moire deflectometry, which directly provides the gradients, is advantageous over interferometric methods.

In this chapter moire deflectometry is introduced and treated mathematically, applying the formalism of indicial equations. To enable a general understanding of the technique, we will overlook, for the moment, the diffraction effects and more subtle aspects involved. Limitations on the sensitivity due to diffraction and imperfection of the optical system will be discussed at the end of the chapter. Throughout the description, an analogy between moire deflectometry and interferometry will be drawn.

6.2. GEOMETRICAL ANALYSIS OF MOIRE DEFLECTOMETRY

A basic moire deflectometer setup consists of a collimated light source and a pair of Ronchi rulings mutually separated in space

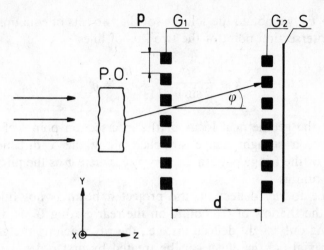

Figure 6.1. The moire deflectometer setup. If the phase object (P.O.) deflects a ray by an angle ϕ, the shadow of the stripes of the grating G_1 will be shifted at G_2 by an amount $d \tan \phi$ where d is the distance between the two gratings. The moire fringe pattern is observed on the screen S.

along the axis of beam propagation. The shadow of the front grating G_1 is projected on the rear grating G_2 as shown in Fig. 6.1. The moire pattern is formed by superimposing the shadow of the front grating with the rear grating and is viewed on a mat screen attached to the rear grating.

In the following analysis of the effect we use purely geometrical optics, thus postponing the treatment of diffraction effects to Section 6.4. We assume that the lines of the front grating G_1 are inclined relative to the x direction at an angle $\theta/2$, whereas those of G_2 are inclined at $-\theta/2$. For a perfectly collimated infinite beam, the shadow of a flawless infinite grating G_1 is described by the set of indicial equations

$$y \cos \theta/2 = x \sin \theta/2 + kp, \qquad k = 0, \pm 1, \pm 2, \pm 3, \ldots , \quad (6.1)$$

and G_2 is described accordingly by

$$y \cos \theta/2 = -x \sin \theta/2 + mp, \qquad m = 0, \pm 1, \pm 2, \pm 3, \ldots . \quad (6.2)$$

Defining $l = k - m$ and using the small angle approximation

$(\sin \theta \sim \theta)$, we obtain the solution of the two sets of equations which is the intersection point of the two sets of lines

$$x = \frac{lp}{2 \sin \theta/2} \cong \frac{lp}{\theta} \equiv lp', \qquad (6.3)$$

namely, the geometrical locus of the intersection points of a given value l is a straight fringe stretched along the x direction. The spacing of the fringe pattern is $p' = p/\theta$, where p is the pitch of the Ronchi rulings.

If, due to ray deflection, the projected beam is not fully collimated, the shadow of G_1 falling on the rear grating G_2 is distorted. Assuming only slight deflections, i.e., the ray reaching the grating is still paraxial (so the effect can be treated by first order, i.e., Gaussian, optics), then ray deflection in the direction of the grating's lines does not affect the moire pattern and can be neglected. On the other hand, a ray deflected at an angle ϕ in the plane perpendicular to the grating's lines, will shift the shadow of G_1 projected on G_2 in the y direction by the amount $d \tan \phi$, where d is the gap between the gratings (see Fig. 6.1).

For small ϕs, we can approximate the latter expression to $d\phi(x, y)$. Hence, Eq. (6.1) will be replaced by the equation describing the distorted shadow of G_1 on G_2, namely,

$$[y + \phi(x, y)d]\cos \theta/2 = x \sin \theta/2 + kp. \qquad (6.4)$$

The solution of the two sets of equations given by Eqs. (6.2) and (6.4) is

$$\phi(x, y)d = 2x \sin \theta/2 + lp. \qquad (6.5)$$

Hence, the deviation δx of the fringe from its unperturbed position [described by Eq. (6.3)] due to ray deflection at an angle $\phi(x, y)$ is given by

$$\delta x \cong \phi(x, y)d/\theta, \qquad (6.6)$$

namely, the fringe shift is linearly proportional to the local deflection angle ϕ, the gap between the gratings d, and the *moire amplification factor* θ^{-1} ($\theta^{-1} = p'/p$ where p' is the pitch of the unperturbed moire pattern). The system parameters are d and θ, however, while an increase of d improves the sensitivity, a change in θ does not; it only contributes to the convenience of the measurement. If $\theta = 0$, we have a configuration known as the *infinite fringe mode*. The equations for the distorted shadow of G_1 and for G_2 are

$$y + \phi(x, y)d = kp,$$

$$y = mp, \tag{6.7}$$

respectively. The solution for this case is

$$\phi(x, y) = lp/d, \tag{6.8}$$

which describes a contour map of ray deflections with increment p/d between adjacent fringes. The same quantity p/d, in the finite fringe mode of operation ($\theta \neq 0$), designates the deflection angle responsible for a unit fringe shift, as derived from Eq. (6.6):

$$\phi(x, y) = \frac{\delta x\, \theta}{d} = \frac{\delta x}{p'}\frac{p}{d}. \tag{6.9}$$

Here we see that the sensitivity of a given setup is proportional to the ratio d/p. The fringe resolution $\delta x/p'$ is physically bounded by the uncertainty principle and is about $1/2\pi$ of a fringe (see Chap. 3). Although the infinite fringe configuration is favored in specific applications such as the study of transient phenomena, the interpretation suffers from a severe drawback in that it indicates only the absolute value of the ray deflection angle and not the direction. A pattern of fringes may result either from a monotonously varying angle of deflection or from fluctuations in that angle, namely, N fringes may represent a maximum deflection angle anywhere between p/d and Np/d.

6.3. RAY DEFLECTION VS. PHASE RETARDATION

In the previous section we introduced moire deflectometry as a tool for ray deflection measurement. It is adequate, therefore, to investigate the causes of ray deflections and to define the relationship between the optical properties of a phase object and its ray deflecting character. Light propagation in space is subject to Fermat's principle, stating that, in going from a given point p_1 to another point p_2, light traverses the route having the smallest optical path length. The mathematical form of this basic principle of ray optics, for an inhomogeneous medium, i.e., a medium with varying refractive index n, is

$$\int_{p_1}^{p_2} n(x, y, z)\, ds = \text{minimum}, \qquad (6.10)$$

where ds is defined as the line element measured along the light ray. The integral expresses the optical path length and the requirement to minimize it is equivalent to minimizing the transit time required for light to travel the distance from p_1 to p_2.

In solving the variational problem presented in Eq. (6.10), Euler equations are used [2]. To bring Eq. (6.10) to an acceptable form for that purpose, ds is transformed to a new variable as

$$ds = \left(dx^2 + dy^2 + dz^2\right)^{1/2} = dz\left(1 + x'^2 + y'^2\right)^{1/2}, \quad (6.11)$$

where $x' = dx/dz$ and $y' = dy/dz$. The optical path length can be rewritten as

$$\int_{p_1}^{p_2} \mathscr{L}(x, y, x', y', z)\, dz = \text{minimum}, \qquad (6.12)$$

where $\mathscr{L}$ (the Lagrangian function) is a function of both x and y and their first derivatives

$$\mathscr{L}(x, y, x', y', z) = n(x, y, z)\left(1 + x'^2 + y'^2\right)^{1/2}. \quad (6.13)$$

The z coordinate is usually chosen to coincide with the optical axis of the system. The Euler equations of the variational problem presented in Eq. (6.12) are

$$\frac{d}{dz}\frac{\partial \mathscr{L}}{\partial x'} - \frac{\partial \mathscr{L}}{\partial x} = 0,$$

$$\frac{d}{dz}\frac{\partial \mathscr{L}}{\partial y'} - \frac{\partial \mathscr{L}}{\partial y} = 0. \qquad (6.14)$$

Substitution of Eq. (6.13) results in

$$\frac{d}{dz}\frac{nx'}{\left(1 + x'^2 + y'^2\right)^{1/2}} = \left(1 + x'^2 + y'^2\right)^{1/2}\frac{\partial n}{\partial x},$$

$$\frac{d}{dz}\frac{ny'}{\left(1 + x'^2 + y'^2\right)^{1/2}} = \left(1 + x'^2 + y'^2\right)^{1/2}\frac{\partial n}{\partial y}, \qquad (6.15)$$

which can be rewritten as

$$\frac{d}{ds}\left(n\frac{dx}{ds}\right) = \frac{\partial n}{\partial x},$$

$$\frac{d}{ds}\left(n\frac{dy}{ds}\right) = \frac{\partial n}{\partial y}. \qquad (6.16)$$

With the help of Eqs. (6.11) and (6.16), $\partial n/\partial z$ can be brought to the same form, namely,

$$\frac{d}{ds}\left(n\frac{dz}{ds}\right) = \frac{\partial n}{\partial z}. \qquad (6.17)$$

Hence, Eqs. (6.16) and (6.17) comprise the three components of the vector ray equation

$$\frac{d}{ds}\left(n(x, y, z)\frac{d\mathbf{r}}{ds}\right) = \nabla n(x, y, z), \qquad (6.18)$$

where $\mathbf{r}$ is the vector drawn from the origin to the light ray and ∇n denotes the gradient of the refractive index.

Using the paraxial approximation, i.e., assuming that the ray deflections ϕ from the optical axis are small ($\sin \phi \approx \tan \phi \approx \phi$ and

$\cos \phi \approx 1$), one can replace ds by dz and rewrite Eq. (6.18) as

$$\frac{d}{dz}\left(n\frac{\partial \mathbf{r}}{\partial z}\right) = \nabla n. \tag{6.19}$$

Now $\partial \mathbf{r}/\partial z$ is simply the ray deflection angle $\phi(z)$ which consists of x and y components, $\phi_x(z)$ and $\phi_y(z)$, respectively. Ray deflection can only be incurred by transverse gradients of the refractive index, i.e., either $\partial n/\partial x$ or $\partial n/\partial y$ should be nonzero. Integrating Eq. (6.19) over the ray path through a phase object of length L yields

$$\phi_x(z) = \frac{1}{n_0}\int_0^L \frac{\partial n}{\partial x}\, dz,$$

$$\phi_y(z) = \frac{1}{n_0}\int_0^L \frac{\partial n}{\partial y}\, dz, \tag{6.20}$$

where n_0 is the ambient refractive index. Thus, the ray deflection by a phase object in the paraxial approximation is the integral of the lateral gradients of the refractive index along the object. Hence, the gradients of the refractive index can be derived from ray deflection measurements obtained by moire deflectometry.

Interferometric methods generally provide the refractive index itself. This value can be derived from moire deflectometry by an additional integration of the quantity measured, performed across the beam in the corresponding direction, namely, over the x axis when $\phi_x(z)$ is measured or along the y coordinate if we have the $\phi_y(z)$ values.

To compare the two methods, a Mach–Zehnder interferometer, which is generally used to measure spatial variations of the refractive index (for example, in compressible gas flows), is presented as an example. As shown in Fig. 6.2, it consists of two beam splitters and two totally reflecting mirrors so that the two waves travel along separate paths. The sample arm includes the phase object to be tested, such as a gas flow chamber, while the reference arm holds an appropriate phase compensator plate. The phase difference between the two waves is given by

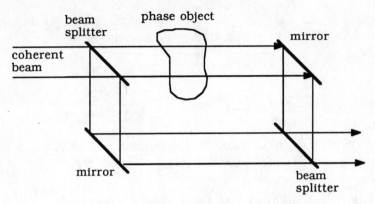

Figure 6.2.   The Mach–Zehnder interferometer.

$$\varphi = \frac{2\pi}{\lambda} \int_0^L [n(x, y, z) - n_0]\, dz. \qquad (6.21)$$

A bright fringe indicates that constructive interference is established, i.e., that the two waves are in phase (namely, $\varphi/2\pi = m$, where $|m| = 0, 1, 2 \ldots$).

Gradients of the refractive index can be measured interferometrically by applying the method known as lateral shearing interferometry. In shearing interferometry, the beam under test is split into two or more laterally displaced beams, which are later brought to mutually overlap with a given lateral displacement which is called the shear. This yields an interferogram of sheared images. If the shear is sufficiently small compared with the spatial resolution desired, then the interference pattern directly corresponds to the spatial derivatives of the optical path. In principle, the inspected phase object can be placed in the path of the single beam or of the multiple beams, produced by the beam splitter. The following discussion assumes the first possibility and $A_1(x, y)$ is the field of a monochromatic plane wave distorted by a phase object,

$$A_1(x, y) = A_0 \exp\left(\frac{-2\pi i z n(x, y)}{\lambda}\right), \qquad (6.22)$$

where λ is the light wavelength in a vacuum and A_0 is the amplitude. The field of a wave sheared by an amount δx in the x direction is given by

$$A_2(x, y) = A_0 \exp\left(\frac{-2\pi i z n(x + \delta x, y)}{\lambda}\right). \quad (6.23)$$

The intensity distribution of the interference pattern formed by the superposition of $A_1(x, y)$ and $A_2(x, y)$ is given by

$$
\begin{aligned}
I(x, y) &= |A_1(x, y) + A_2(x, y)|^2 \\
&= 2A_0^2\left(1 + \cos\frac{2\pi z}{\lambda}[n(x + \delta x, y) - n(x, y)]\right). \quad (6.24)
\end{aligned}
$$

When the refractive index varies slowly and the shear δx is sufficiently small, then $I(x, y)$ is approximated by

$$I(x, y) = 2A_0^2\left(1 + \cos\frac{2\pi z}{\lambda}\,\delta x\left[\frac{\partial n(x, y)}{\partial x}\right]\right). \quad (6.25)$$

In the general case, the refractive index is also a function of z, the path traveled along the phase object of length L. Equation (6.25) should be rewritten as

$$I(x, y) = 2A_0^2\left(1 + \cos\frac{2\pi}{\lambda}\,\delta x\int_0^L \frac{\partial n(x, y, z)}{\partial x}\,dz\right). \quad (6.26)$$

The condition for constructive interference, i.e., for bright fringe to occur, is

$$\frac{2\pi}{\lambda}\,\delta x\int_0^L \frac{\partial n(x, y, z)}{\partial x}\,dz = 2\pi k, \qquad k = 0, \pm 1, \pm 2, \dots. \quad (6.27)$$

To be consistent with the analogy to moire deflectometry, Eq. (6.27) is rearranged to be

$$\frac{1}{n_0} \int_0^L \frac{\partial n}{\partial x} \, dz = \frac{k\lambda}{n_0 \, \delta x}. \tag{6.28}$$

This expression, in which the left-hand-side is equal to $\phi_x(x, y)$, resembles the one obtained in moire deflectometry for the infinite fringe configuration ($\theta = 0$), given in Eq. (6.8).

The infinite fringe moire deflectogram is a contour map of ray deflections incremented by p/d. The full analogy between shearing interferometry and moire deflectometry is demonstrated by rederiving Eq. (6.8) via physical optics. Let us regard the moire deflectometer as a shearing interferometer, using the grating G_1 as the beam splitter. By using a Ronchi ruling as the beam splitter, the beam is divided into numerous diffraction orders of which only the first ones ($+1, 0$, and -1), which contain over 80% of the diffracted light intensity, are considered. The shear between two adjacent images is given by the diffraction angle. Rigorous treatment of shearing interferometry based on diffraction theory has been carried out by Yokozeki and Suzuki [3] and Lohmann and Silva [4]. We adopt a shorthand formulation. An infinite Ronchi ruling with pitch p aligned in the x direction can be expressed by $\psi(x)$,

$$\psi(x) = \text{rect}\left(\frac{2x}{p}\right) * \text{comb}\left(\frac{x}{p}\right), \tag{6.29}$$

where $*$ denotes a convolution operation,

$$\text{rect}\left(\frac{2x}{p}\right) = \begin{cases} 1 & |x| \le p/4, \\ 0 & \text{otherwise} \end{cases} \tag{6.30}$$

and

$$\text{comb}\left(\frac{x}{p}\right) = \sum_{k=-\infty}^{\infty} \delta(x - kp), \tag{6.31}$$

namely, the infinite Ronchi ruling is described by a convolution of a rectangle function with a comb function. Assuming far field (Fraunhofer) diffraction, the diffraction field at a distance z from the grating is given by the Fourier transform of the grating function $\psi(x)$, multiplied by phase factors denoted by K, which are independent of x, namely,

$$\psi(x') = K \int_{-\infty}^{\infty} \psi(x) \exp\left(-\frac{ixx'n_0}{\lambda z}\right) dx, \qquad (6.32)$$

where x' denotes the x-axis coordinate on a plane displaced by z from the grating's plane. By substituting $\psi(x)$ in Eq. (6.32), we obtain

$$\psi(x') = K \operatorname{sinc}\left(\frac{px'}{2\lambda z}\right) \sum_{k=-\infty}^{\infty} \delta\left(\frac{x'n_0}{\lambda z} - \frac{k}{p}\right). \qquad (6.33)$$

The contribution of the multiple slit to diffraction is expressed by the sum term of Eq. (6.33), so that the diffraction angle for the kth diffraction order can be found by assuming nonzero values for the δ function, namely, $x' = (kz\lambda/pn_0)$ where x' expresses the location of the kth order of diffraction at a distance z from the diffraction grating. For a small diffraction angle α we can approximate

$$\alpha \cong \frac{x'}{z} = \frac{k\lambda}{pn_0}. \qquad (6.34)$$

Since most of the light energy is concentrated in the lowest diffraction orders (0 and ± 1), only the first order diffraction angle (λ/pn_0) will be considered. Using diffraction gratings as a shearing device, the shear between two images of a beam at a distance d is $\delta x = (\lambda d/pn_0)$.

We now demonstrate how a deflectometer acts as a shearing interferometer using the two grating arrangement of Fig. 6.3. For simplicity, only the zeroth and first diffraction orders are shown. The two numbers designate the diffraction orders of the first and second gratings, respectively, and therefore, for example, the beam 10 is the first order beam of the grating G_1 and the zeroth order beam of G_2. It is seen that the two beams 01 and 10 have the same direction in space and thus interfere, although they are displaced by the shear δx.

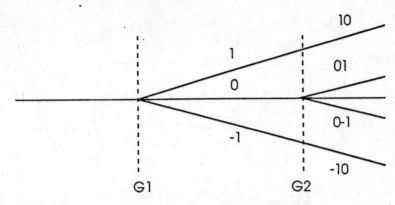

Figure 6.3. The moire deflectometer as a shearing interferometer. G_1 and G_2 represent the two gratings and the numbers represent the diffraction orders of the first and second grating, respectively.

Similarly, $0-1$ and -10 will also interfere and a multiple beam shearing interferometer is formed. Substituting δx in Eq. (6.28) we obtain

$$\frac{1}{n_0} \int_0^L \frac{\partial n}{\partial x}\, dz = \frac{kp}{d},$$
(6.35)

which is identical with the expression obtained from geometrical optics and Fermat's principle (i.e., light rays propagate in a trajectory that minimizes the optical path length).

It is appropriate to mention that although the discussion so far dealt with phase objects in which ray deflections are caused by lateral derivatives of the refractive index, the same conclusions also hold for specular objects, only with a minor change. In the case of reflection from a surface where rays propagate all the time in a homogeneous medium, the requirement to minimize the optical path is replaced by a simpler expression

$$\int_{P_1}^{P_2} ds = \text{minimum},$$
(6.36)

namely, that the geometrical path itself should be minimum. In the paraxial approximation, the deflection angle ϕ due to deviation from flatness, i.e., a slope of $\partial h/\partial x \cong \beta$ (where h denotes the surface elevation) is

$$\phi = 2\beta. \tag{6.37}$$

The same analogy with interferometry can be derived for moire deflectometry of surfaces.

6.4. SENSITIVITY OF MOIRE DEFLECTOMETRY

The accuracy of moire deflectometry is determined, among other factors, by the fringe resolution, i.e., the minimum discernible fringe shift. The fringe profile, obtained by the superposition of two perfect Ronchi rulings brought to physical contact, is triangular (see Chap. 1). However, as the gratings are set apart from each other, the idealized fringe pattern deteriorates and blurs due to several factors such as poor beam quality, imperfection of the gratings, and inevitable diffraction due to the finite aperture. Therefore, the accuracy of the measurement is determined by the quality of the contrast of the fringe pattern. Generally speaking, the naked eye can resolve a shift of about one-tenth of a fringe. Measuring the intensity of the light transmitted through a set of gratings can be automated using light detectors coupled to a readout device, but this does not increase the fringe resolution significantly. The techniques that can be used to increase the fringe resolution are discussed in Chap. 2.

A shift of one fringe corresponds to a deflection angle $\phi = p/d$, i.e., the sensitivity of the measurement is determined by the ratio p/d. As we reduce p/d to improve the angular sensitivity, the fringe pattern is blurred due to diffraction caused by both the diffusiveness of the medium and finite aperture of the optical system. The ultimate sensitivity is determined by the diffraction limit of the system.

6.5. DIFFRACTION BY A GRATING—TALBOT EFFECT

We now discuss the various effects that limit the sensitivity of the method in a simple, merely physical, manner. When a collimated

beam of coherent light falls upon a transmission grating, it is split
into numerous diffracted waves propagating at angles $k\lambda/p$, where k
is the diffraction order. Since the waves are mutually coherent they
interfere among themselves. Let us assume that an infinite flawless
grating is located at a distance d, where the lateral displacement of
the beam of the kth diffraction order equals the grating's period p,
namely,

$$\frac{\lambda d}{p} = kp. \qquad (6.38)$$

All the diffraction orders are located in the same way as on the
grating ($d = 0$) and therefore a constructive interference between all
the diffraction orders of the arbitrary grating is established. The
interference pattern is an exact replica of the object grating. The
distance $d_T = p^2/\lambda$ is known as the *Talbot self-imaging plane* (or
Fourier plane) [5] after Talbot who published his discovery of the
self-imaging effect of gratings in 1876 [6]. At distances that are
integer multiples of d_T, the blur owing to the grating's diffraction is
minimized, i.e., the contrast of the fringe pattern attains a local
maximum. Therefore, by adjusting the gap between the Ronchi
rulings to integer multiples of d_T, we can avoid diffraction by the
gratings and justify our approach to moire deflectometry as a merely
geometrical optics phenomenon.

It should be emphasized, however, that well-distinguished planar
Talbot images are only obtained for an ideally collimated infinite
beam. Any ray deflecting element will blur the pattern to some
extent or change the location of the planes. Moreover, the interpretation suggested here, which assumes Fraunhofer (far field) diffraction,
is oversimplified. A rigorous treatment based on the more justifiable
model of Fresnel diffraction, which also explains other more subtle
aspects of diffraction, such as the change in sign of the fringes
between odd and even Talbot planes, is given in ref. 5.

6.6. DIFFRACTION BY A FINITE APERTURE

Referring again to the diffraction effect of gratings, one should
notice that the lateral separation between the first and zero orders of

diffraction, the shear, cannot exceed the size of the effective aperture a, namely,

$$d_{max}\frac{\lambda}{p} \leq a \qquad (6.39)$$

or, alternatively,

$$k_{max} \leq \frac{a}{p}. \qquad (6.40)$$

Therefore, the minimum detectable deflection angle is given by

$$\phi_{min} = q\frac{p}{d_{max}} = q\frac{\lambda}{a}, \qquad (6.41)$$

where q is the fringe resolution, i.e., the minimum discernible fringe shift in terms of fringe fraction. The quantity λ/a is the expression for diffraction-limited divergence, a general concept in optics, and is not specific to our measurement. It is actually the optical form of the Heisenberg uncertainty principle, a cornerstone of wave mechanics. This principle states that the angular and spatial resolution are linked together, so that their product cannot be smaller than $\lambda/2\pi$. Therefore, improving the angular sensitivity by increasing the gap between the gratings, inevitably results in reduced spatial resolution. Since the lower bound to spatial resolution is the effective aperture, the angular sensitivity cannot be improved beyond the diffraction-limited divergence (see Chap. 3).

6.7. THE EFFECT OF BEAM QUALITY

The divergence of a perfectly collimated beam is determined by the beam width, as just discussed, and is referred to as the diffraction-limited divergence. One should distinguish between reversible divergence, which can be corrected by passive optical components (like lenses), and irreversible divergence. Irreversible divergence that exceeds the diffraction limit is related to reduced beam quality incurred upon passing through diffusive media. The effect of this excessive divergence is to reduce the fringe contrast, or modulation, which is given by

$$M = \frac{I_{max} - I_{min}}{I_{max} + I_{min}}, \tag{6.42}$$

where I_{max} and I_{min} denote the maximum and minimum brightness of the moire pattern. The extent to which the modulation decreases on passing an optical system is expressed by the modulation transfer function (MTF) of that system. The MTF is determined as the ratio of the modulation of the image produced by the system and the modulation of the object, and shows a strong dependence on the spatial frequency. Although the modulation and MTF are defined in that way for sinusoidally varying patterns, they can be applied to moire patterns, introducing a small error due to contributions of Fourier series terms of orders higher than 1.

The MTF of a system can be related to the quantum statistical concept of *transverse modes of radiation* [8] (see Chap. 3). Intuitively, the number of transverse modes n of a monochromatic beam of width a, wavelength λ, and solid angle divergence $\delta\Omega$, is approximately

$$n \cong \frac{\delta\Omega}{(\lambda/a)^2}. \tag{6.43}$$

Here, λ/a is the 1-D diffraction-limited divergence or the divergence of a single-mode beam, assuming that each separate mode occupies an equal section of the angle space. We also claim that each spatial mode represents a wave traveling in a certain direction and $\delta\Omega$ is the characteristic width of the angular distribution of the waves $B(\alpha)$, where the variable α denotes the scattering angle. For simplicity, we shall assume uniform divergence within the solid angle $\delta\Omega$, namely,

$$B(\alpha) = \text{rect}\left(\frac{\alpha}{(\delta\Omega)^{1/2}}\right). \tag{6.44}$$

The fringe shape of a moire deflectogram affected by that divergence is no longer triangular. It can be described as the convolution of the triangle function with the angular distribution function $B(\alpha)$. The triangular function (see Chap. 1) is given by

$$T(x) = \frac{2(I_{max} + I_{min})}{p'}|x|, \qquad |x| \le \frac{p'}{2}, \qquad (6.45)$$

while the width of the rectangular function is $[(\delta\Omega)^{1/2} dp'/p]$, which is assumed to be much smaller than p'. The convolution around the bottom of the fringe yields the intensity distribution $I(x')$, where

$$I(x') = \frac{p(I_{max} + I_{min})}{(\delta\Omega)^{1/2} dp'^2} \int_{x' - ((\delta\Omega)^{1/2} dp')/2p}^{x' + ((\delta\Omega)^{1/2} dp')/2p} |x| \, dx$$

$$= \frac{2p(I_{max} + I_{min})}{(\delta\Omega)^{1/2} dp'^2} \left[x'^2 + \left(\frac{(\delta\Omega)^{1/2} dp'}{2p} \right)^2 \right]. \qquad (6.46)$$

Hence,

$$I_{min} = I(0) = \frac{I_{max}(\delta\Omega)^{1/2} d}{2p - (\delta\Omega)^{1/2} d} \qquad (6.47)$$

and

$$M = 1 - \frac{(\delta\Omega)^{1/2} d}{p} = 1 - \frac{n^{1/2}\lambda d}{pa}. \qquad (6.48)$$

The expression $\lambda d/pa$ is equal to the reduced spatial frequency s [7], defined as the ratio of the absolute spatial frequency ($1/p$ in this case) and the cutoff frequency determined by the size of the diffraction-limited image of a point source ($\sim d\lambda/a$). Thus the MTF of a square aperture with a square angular distribution can be approximated as

$$M(s) = 1 - n^{1/2}s. \qquad (6.49)$$

6.8. THE EFFECT OF GRATING IMPERFECTION

In introducing the Talbot effect of self-imaging by gratings, an infinitely large and flawless diffraction grating was considered. The effect of limiting aperture was already discussed and now we will refer to the grating quality. Since the self-imaging is based on the interference between light diffracted from spatially separated slits of the grating, a perfect replica of the object grating will be formed only if the grating's periodicity is strictly retained across the grating. This requirement is the spatial analog to the demand for strictly monochromatic light to maintain temporal coherence. The temporal coherence of light is characterized by the *coherence length*, which is proportional to the reciprocal of the spectral bandwidth. This relation satisfies the uncertainty principle (see Chap. 3)

$$\delta t \, \delta \nu \geq \frac{1}{2\pi}, \tag{6.50}$$

where $\delta t = \delta(L/c)$ (c is the light velocity and L is the geometric path length). We shall develop an expression, analogous to the coherence length, for the spatial coherence of the grating. The coherence length is defined by the number of wavelengths that the light has to propagate until it completely loses its phase information. In analogy we can relate to the distance in grating periods required to bring the stripes of the gratings out of phase. We shall again use Eq. (6.29) to describe an infinite perfect Ronchi ruling:

$$\psi_p = \text{rect}\left(\frac{2x}{p}\right) * \sum_{k=-\infty}^{\infty} \delta(x - kp). \tag{6.29'}$$

A realistic Ronchi ruling, can be defined by a general expression

$$\psi_r = \sum_{k=-N/2}^{N/2} \text{rect}\left[\frac{2(x - x_k)}{p}\right]. \tag{6.51}$$

In the case where $x_k = kp$ and $N \Rightarrow \infty$, then $\psi_r = \psi_p$. To find the *degree of coherence* of the grating we can apply the mutual correlation function $\Gamma_{1,2}(x)$, defined as the mixed term in the averaged intensity of the superposition of wave functions ψ_1 and ψ_2, namely,

$$\Gamma_{1,2}(x) = \int_{-\infty}^{\infty} \psi_1(x')\psi_2(x' + x)\, dx' \equiv \langle \psi_1 \psi_2 \rangle. \quad (6.52)$$

To find the coherence length of a real grating ψ_r, we might cross-correlate it with itself:

$$\Gamma_{rr}(x) = \langle \psi_r(x')\psi_r(x' + x) \rangle. \quad (6.53)$$

The function can be normalized to yield the *complex degree of coherence* γ_{rr}:

$$\gamma_{rr}(x) = \frac{\Gamma_{rr}(x)}{\Gamma_{rr}(0)} = \frac{\langle \psi_r(x')\psi_r(x' + x) \rangle}{\langle |\psi_r|^2 \rangle} \quad (6.54)$$

If $\psi_r = \psi_p$, the modulus of the function $|\gamma_{rr}(x)|$ is unity, i.e., the coherence is complete and the fringe contrast will not be additionally reduced due to the gratings' quality. On the other hand, if the grating's kth slit is not located at the exact distance $x = kp$ from the 0 indexed slit, but, for example, at $x = (k + 1/2)p$, then at the kth Talbot plane (i.e., $d = kp^2/\lambda$) the fringe pattern will be totally blurred. This is due to the destructive interference between the zero order diffraction of the kth slit and the first order diffraction of the zero indexed slit of the grating, which are out of phase.

Hence, we can express the *spatial coherence length* of a Ronchi ruling as mp, where m is the index of the Talbot plane on which the fringe modulation falls to $1/e$ from its contrast at contact. The physical meaning is that the grating loses the phase information over a distance of k periods. The coherence length can be measured using a highly collimated (single transverse mode) broad beam and observing the contrast reduction away from the aperture edge as d increases until the fringes are smeared out completely. For perfect gratings the fringe pattern will vanish when $kp = a$ (a is the limiting aperture), as discussed previously.

REFERENCES

1. O. Kafri, *Opt. Lett.* **5**, 555 (1980).
2. I. M. Gelfand and S. V. Fonin, *Calculus of Variations*, Prentice Hall, Englewood Cliffs, N.J., 1963.
3. S. Yokozeki and T. Suzuki, *Appl. Opt.* **10**, 1575 (1971).
4. A. W. Lohmann and D. E. Silva, *Optics Commun.* **2**, 1690 (1971).
5. E. Keren and O. Kafri, *J. Opt. Soc. Am.* **A2**, 111 (1985).
6. H. Talbot, *Philos. Mag.* **9**, 401 (1876).
7. J. W. Goodman, *Introduction to Fourier Optics*, McGraw-Hill, N.Y., 1968.
8. Z. Karny, S. Lavi, and O. Kafri, *Opt. Lett.* **8**, 409 (1983).

7

Applications of Moire Deflectometry

7.1. INTRODUCTION

In the previous chapter we introduced moire deflectometry, a method for measuring ray deflections. We showed how both directional and random (light scattering) ray deflections affect the moire fringe pattern. A thorough correlation between the fringe shift and the deflection angle, and between the fringe contrast and the beam quality (or the MTF of a test object) was drawn. In this chapter we demonstrate and give examples of how the optical properties of objects and transparent media are determined by moire deflectometry.

We first discuss the different types of information that can be derived from the fringe shift. In general, a directional ray deflection is caused either by a phase distorting optical object or is due to changes in the relative inclination angle of an object in space. The main use of moire deflectometry is in the general testing of optical objects, either static or in dynamic systems. The static objects are optical elements or systems, transparent and specular alike, whose optical character does not change in time. In this category we include all routine tests of optical components. The dynamic systems studied by moire deflectometry comprise a variety of physical phenomena that affect the course of rays, thereby leading to ray deflection or scattering of light. Although the optical properties studied in the dynamic systems are in principle the same as in static measurements, the rapid variations in time require fast recording of the signal provided by the deflectometer, and therefore these measurements deserve separate treatment in this text. Examples of dynamic systems studied by moire deflectometry include wind tunnels, fluid mixtures, and nonuniformly irradiated absorbing media.

111

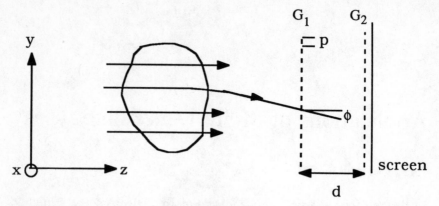

Figure 7.1. The basic moire deflectometer setup. A paraxial beam projects a shadow of the first grating G_1 onto the grating G_2 separated by a distance d.

7.2. EXPERIMENTAL TECHNIQUES IN MOIRE DEFLECTOMETRY

To answer the specific demands of the different applications of moire deflectometry, the basic setup shown in Fig. 7.1, which comprises a collimated light source, a pair of matched Ronchi rulings, and a mat screen for visual display of the fringe pattern, has to be modified. For small scale objects, a microscopic-type moire deflectometer was constructed [1]. To increase the angular resolution while maintaining the compact structure, a folded telescopic device was designed [2]. Different techniques of fringe reading, both in real time and for deferred analysis, were devised to improve the accuracy of the measurement and bring it close to the theoretical limit. We describe the various modifications of moire deflectometry currently in use.

7.2.1. Configurations of the Moire Deflectometer

7.2.1.1. Reflective Surfaces. The basic setup of Fig. 7.1 can be used only for transparent objects. Reflecting surfaces require a slight change in design, as shown in Fig. 7.2. The collimated beam is projected at an oblique incidence angle on the surface to be tested and the reflected beam is then directed at normal incidence on the

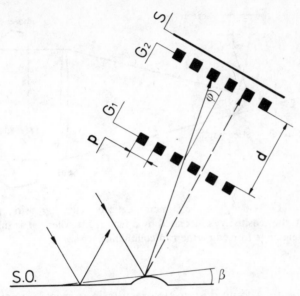

Figure 7.2. Moire deflectometry in the reflection mode. A ray deflection $\phi = 2\beta$ results from a local slope β on the object.

gratings. An elevation on the inspected surface with a slope angle β will deflect the ray away from normal incidence by $\phi = 2\beta$ [3].

7.2.1.2. Folded Telescopic Setup. An improved setup, which is applicable for both phase objects and specular surfaces, is offered by a telescopic configuration shown in Fig. 7.3 [2]. The collimated beam is produced by a beam expander and a large objective lens. It passes twice through the phase object because of a flat mirror positioned behind the object. The retraced beam is diverted at a right angle by a beam splitter into a smaller objective lens, which recollimates the beam on the small diameter deflectometer. A flat specular object is tested by removing the flat end mirror and replacing it by the test object. Testing of a mirror by illumination at normal incidence, instead of the oblique incidence offered by the setup of Fig. 7.2, eliminates the inherent distortion of cos α (α is the incidence angle) as well as the shadowing effects occurring at high incidence angle.

Other advantages of this setup are improvement in mechanical stability and reduced susceptibility to aberrations in the collimating

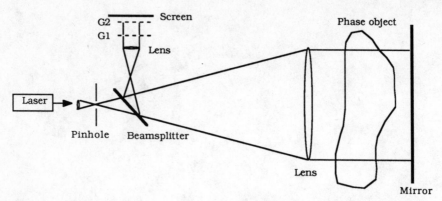

Figure 7.3. The folded telescopic moire deflectometer. By using a telescopic arrangement, the sensitivity is increased by a factor $2M$ where M is the magnification of the telescope formed by the two collimating lenses.

system. The telescopic design also results in increased angular sensitivity. If the magnification (the ratio of the focal lengths of the two objective lenses) is M, a ray deflected by ϕ at the test object will enter the deflectometer at an angle $2M\phi$, as can be proved by geometrical ray tracing.

The advantage of the magnified angular resolution is that high sensitivities can be reached by the telescope arrangement at a gratings' gap smaller by a factor $2M$ compared to the linear setup. As a result, the noise due to imperfections of the gratings and the blur caused by diffraction will be reduced by a factor $2M$. However, as implied from the uncertainty principle, an increase in the angular resolution by a factor M is followed by reduction in the spatial resolution by the same factor. Therefore, the telescopic setup is mainly intended for large scale objects of relatively low deformation in which a moderate spatial resolution is required (for example, atmospheric measurements and flow field visualization in wind tunnels). An example of a telescopic moire deflectometer measurement [4] is shown in Fig. 7.4. Two deflectograms of a candle's flame, one made in the basic setup with angular resolution of 3×10^{-4} rad (Fig. 7.4a) and the other (Fig. 7.4b) taken using the telescope assembly of magnification $7\times$, demonstrate the increase in angular sensitivity.

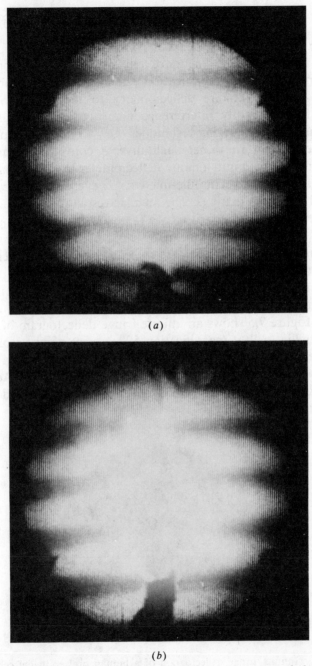

Figure 7.4. A candle flame (*a*) in the low resolution basic setup and (*b*) with a 7×
magnification telescopic setup.

7.2.1.3. Microscopic Setup. The idea of a telescope applied in a reverse fashion led to the development of a microscopic moire deflectometer [1]. This modification extends the range of objects that can be tested by moire deflectometry to the submicron region. Except for the costly interference microscope, other microscopic phase distortion analyzing techniques (i.e., schlieren and phase contrast microscope) are rather qualitative. A conventional microscope can be transformed into a moire deflectometer by using a laser as the light source and replacing the microscope eyepiece by a collimating lens and a pair of gratings. The angular resolution of a microscopic moire deflectometer decreases by a factor M (the microscope magnification). Since most of the microscopic objects of interest deflect light quite strongly, the decrease in sensitivity can be tolerated. On the other hand, the gain in spatial resolution by the same magnification factor extends the applicability of moire deflectometry far beyond the limit set by the gratings pitch (usually about 25 μm or higher). Figure 7.5 shows an infinite fringe deflectogram of a human oral epithelial cell at magnification $500\times$. For comparison a reference microphotograph of the same cell was obtained by a standard microscope with a laser illuminator, as shown in Fig. 7.6. Adaptation to testing of microscopic reflecting surfaces can be achieved easily.

Figure 7.5. Infinite fringe deflectogram of a human oral epithelial cell at $500\times$ magnification.

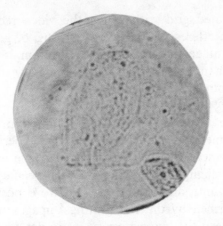

Figure 7.6.   Reference microphotograph of the object of Fig. 7.5 obtained using a regular microscope with a laser illuminator.

7.2.2.   Real Time vs. Deferred Analysis Methods

7.2.2.1.   Deferred Analysis. Although real time observation is the common practice in moire deflectometry, storing the deflectogram, either photographically or in a computer memory, offers a variety of useful operation modes. The advantage of deferred analysis over real time measurements is obvious in the recording of transient phenomena with short lifetimes. Once the data are stored they can be further processed in various ways to optimally extract valuable information encoded in the moire pattern. One way of conserving a transient effect is by freezing the distorted beam emerging from a phase object, or reflected from a mirror, for post-analysis. The method suggested by Stricker and Politch is holography [5]. A hologram of the phase object is recorded on a photographic plate. The real image of the beam diffracted from the hologram is later reconstructed and analyzed by a conventional moire deflectometer.

Deferred moire deflectometric analysis is generally performed by replacing the second (reference) grating with photographic film. The transparency of the distorted grating image at a preselected gap can be superimposed with the reference grating to obtain a fringe pattern at any desired intersection angle [6]. The second grating can be a computer generated grid with which the video recorded distorted grating image interferes. The distorted grating, or grating hologram, has the disadvantage that the spatial resolution is dictated by the grating rather than by the moire fringes, and we therefore do not benefit much from the moire analysis.

7.2.2.2. Double Exposure. The most common practice of single grating deflectometry is for double exposure experiments. For this type of measurement the distorted grating image is recorded twice while a certain change, depending on the specific purpose of the measurement, is introduced between the two exposures. The moire deflectogram formed by the two gratings' images reveals hyperfine differences between the two exposures, as in double exposure holography. Double exposure can be applied for various purposes, such as removing distortions caused by imperfections of the optical components of the measuring system [7]. A picture of the system without the test object is taken. The object is then inserted, the grating rotated by an angle θ, and the film is exposed again. In the resulting moire deflectogram, shown in Fig. 7.7, the phase distortion caused by the system itself is removed. The phase object in this example is hot air produced by a soldering iron. The phase distorting element intentionally placed in both exposures was a twisted Perspex plate. The double exposure correcting effect is explained by the two indicial equations describing the images projected at a distance d from the grating, as obtained for the two inclination angles $\theta/2$ and $-\theta/2$ relative to the x axis. To simplify, let us assume that both images are distorted by the same deflection angle ϕ:

$$(y + \phi d)\cos \theta/2 = x \sin \theta/2 + kp, \qquad k = 0, \pm 1, \pm 2, \ldots,$$
$$(y + \phi d)\cos \theta/2 = -x \sin \theta/2 + mp, \qquad m = 0, \pm 1, \pm 2, \ldots.$$
$$(7.1)$$

The solution is a set of straight fringes along the y direction, as obtained for the undistorted case where $\phi = 0$. The sensitivity of this method is limited by the grating density (which is limited by the viewing system) and not by the moire fringes.

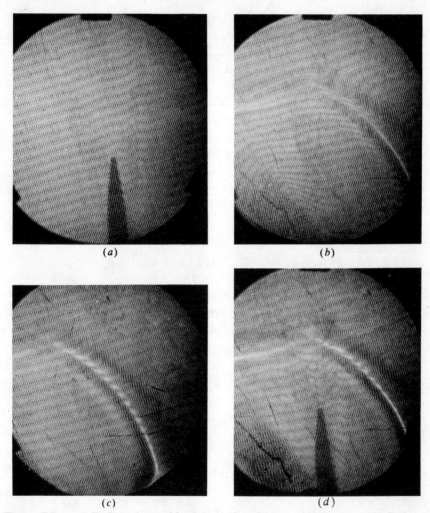

Figure 7.7. (*a*) Deflectogram of hot air produced by a soldering iron. (*b*) Deflectogram of a distorted Perspex plate. (*c*) Double exposure deflectogram of the Perspex plate of Fig. 7.7*b*. (*d*) Double exposure deflectogram of the hot air observed through the Perspex plate. Note that the noise from the Perspex has been removed.

7.2.2.3. Shearing Moire Deflectometry. Another important application of the double exposure method is for shearing moire deflectometry [8], i.e., the measurement of the lateral gradients of the deflection angle. This is equivalent to the second derivative of the refractive index in phase objects or the curvature in the case of surface reflection. The measurement is accomplished in the same setup as the correcting double exposure, except that the test object is present in both exposures. The shearing is performed by slightly shifting the object by δy in the y direction between the two exposures. To obtain a finite fringe pattern the grating is also rotated by θ between the two exposures. The indicial equations for the two images are

$$[y + \phi(x, y)d]\cos \theta/2 = x \sin \theta/2 + kp,$$
$$[y + \phi(x, y + \delta y)d + \delta y]\cos \theta/2 = -x \sin \theta/2 + mp. \quad (7.2)$$

The moire pattern obtained by the double exposure is distorted according to the difference between $\phi(x, y)$ and $\phi(x, y + \delta y)$. For a small shift this is proportional to the derivative of ϕ, namely,

$$\frac{x'\theta}{d\,\delta y} = \frac{\phi(x, y) - \phi(x, y + \delta y)}{\delta y} \cong \frac{\partial \phi}{\partial y}, \quad (7.3)$$

where x' is the fringe distortion [see Eq. (2.43)]. The interpretation of Eq. (7.3) for phase objects is the following: Since,

$$\phi = n_f^{-1} \int_0^L \left(\frac{\partial n}{\partial y} \right) dz,$$

where n_f is the refractive index at the exit from the phase object and L is the object length, then

$$\frac{\partial \phi}{\partial y} \cong \frac{1}{n_f} \int_0^L \frac{\partial^2 n}{\partial y^2} \, dz. \quad (7.4)$$

The sheared deflectogram thus yields the second derivative of the

refractive index of a phase object. For reflection from a test surface
$\phi \cong 2(\partial h/\partial y)$, where h is the surface elevation, we obtain

$$\frac{\partial^2 h}{\partial y^2} \cong \frac{x'\theta}{2d\,\delta y}, \qquad (7.5)$$

namely, a deviation of the fringes from linearity x' is proportional to
the curvature. Figure 7.8 demonstrates the application of shearing
moire deflectometry to a curved surface.

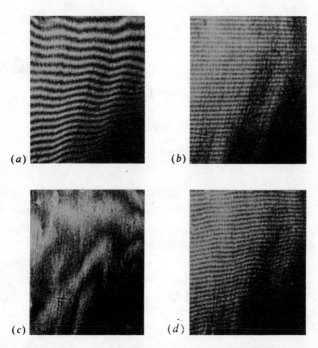

Figure 7.8. Shearing by moire deflectometry. (*a*) Simple deflectogram of a curved
surface. The fringe distortion is proportional to the first height derivative. (*b*)
Double exposure moire deflectogram of the distorted grating without shift. Note
that the fringes are straight. (*c*) Double exposure deflectogram in infinite fringe
mode. The object was shifted between exposures and the contours are lines of
equisecond derivatives. (*d*) Sheared double exposure deflectogram in finite fringe
mode. The fringe distortion is proportional to the second height derivative.

The shearing process can be repeated to obtain higher-order derivatives [9]. The third derivative of the elevation ($\partial^3 h/\partial y^3$) is achieved by superimposing a finite fringe moire deflectogram with its reproduction, which introduces a small shift δy_1 between the two. If the moire pattern is described by the indicial equation

$$x = G(y, x) + kp', \tag{7.6}$$

where $p' \cong p/\theta$, then the *moire of moire* pattern of the shifted deflectogram will yield

$$\frac{\partial G}{\partial y} \cong \frac{G(y, x) - G(y + \delta y_1, x)}{\delta y_1} = \frac{lp'}{\delta y_1}, \tag{7.7}$$

where δy_1 denotes the shift of the deflectogram. A contour map of the third height derivative is given by [see Eq. (2.49)]

$$\frac{\partial^3 h}{\partial y^3} \cong \frac{lp}{2d\,\delta y\,\delta y_1}. \tag{7.8}$$

The superposition of two moire deflectograms to obtain a secondary moire pattern (moire of moire) is useful for the measurement of small fringe rotation angles ($\alpha < 10^{-2}$ rad) caused by lenses or spherical mirrors [10]. In this application the two moire patterns, one of which is a reference while the other is taken with the test object, are brought to overlap [11]. The appearance of a secondary moire pattern of pitch p'' reveals the fringe rotation. From this pitch one can calculate α with high accuracy as $\alpha = 2\sin^{-1}(p'/2p'')$.

7.2.2.4. Additive Moire Deflectometry. Superposition of moire deflectograms offers yet another application in the summation of distortions, rather than the mutual subtraction already demonstrated. This option can be used in optical design for predicting the ray course on passing a series of known components [12]. To obtain fringe addition, conjugate moire patterns have to be superimposed.

Two gratings are conjugate (in moire deflectometry) if their indicial equations are of the form

$$y + f(x, y) = kp,$$
$$y - f(x, y) = mp. \tag{7.9}$$

A conjugate pair of moire patterns consists of a moire deflectogram, in which the reference grating is rotated by θ relative to the front grating, and a second deflectogram, in which the reference grating is rotated by $-\theta$. For ray deflections ϕ_1 and ϕ_2 introduced into the two conjugate deflectograms, respectively, the indicial equations would be (see 2.2.3)

$$x = \frac{kp}{\theta} + \frac{\phi_1 d}{\theta},$$
$$x = \frac{mp}{\theta} - \frac{\phi_2 d}{\theta}, \tag{7.10}$$

for which the solution is

$$\phi_1(y, x) + \phi_2(y, x) = \frac{lp}{d}, \tag{7.11}$$

namely, an infinite fringe moire pattern of the combined deflections.

7.3. APPLICATIONS OF MOIRE DEFLECTOMETRY IN STATIC SYSTEMS

The applications of moire deflectometry to ray deflection analysis are classified as: (1) static systems, in which we include the testing of optical elements; (2) dynamic systems, which includes flow systems and transient phenomena; (3) miscellaneous applications.

For static systems we analyze the basic types of ray deflections on the fringe pattern and show some applications in the characterization of optical components. We distinguish between: (1) deflection of the entire beam at a common angle, caused by a constant gradient; (2) deflection by a linearly increasing angle, caused by a parabolical mirror or an object with a parabolically varying refractive index; (3) deflection from phase distorting objects in general. The analysis of the first two types is straightforward. The first causes merely a parallel displacement of the entire fringe pattern, while the second is observed as a rotation of the fringe pattern about an axis of symmetry. Ray deflections of the third type cause local fringe shifts and the interpretation usually requires numerical analysis.

7.3.1. Beam Deflecting Objects

Deflection of a beam from its normal incidence on the deflectometer is caused either by a specular object or by a wedge-shaped prism. Figure 7.9 demonstrates ray deflection by a wedge with an inclination of $(\alpha + \beta)$ between the two end surfaces [13]. Assuming that the wedge angle is small, the deflection angle ϕ is given by $(n - 1)(\alpha + \beta)$, where n is the refractive index of the wedge. For a constant ϕ along the y axis of the gratings, the pattern deformation is a parallel translation of the fringes. The wedge angle is found from the fringe displacement x' with respect to its unperturbed position,

$$\alpha + \beta = \frac{px'}{d(n - 1)p'}. \tag{7.12}$$

Since the deflectometer is sensitive to ray deflections along the y axis only, the wedge has to be rotated about its center until the maximum shift is observed. By further rotating by 180° the fringe is shifted to the maximum in the opposite direction. It is convenient to derive the

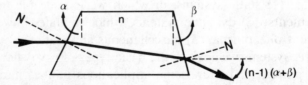

Figure 7.9. Ray deflection of a collimated beam passing through a wedged window of refractive index n.

wedge angle from half the difference in fringe shift in both directions upon rotation by 180°. The sensitivity in the measurement of wedge angle, or the degree of parallelism of the end surfaces, is limited by the minimum detectable fringe shift. For small objects the limitation imposed by the uncertainty principle should also be considered, i.e., the angular resolution cannot be better than $\lambda/4\pi\,\delta x$, where δx is the spatial resolution required. Measurement of the wedge angle by transmission, as done by moire deflectometry, is superior to an autocollimator. The autocollimator measures the inclination of each surface separately via reflection, sometimes very weak, and compares the results obtained for the two surfaces, thus doubling the error. In addition, the deflectometer is practically insensitive to the absolute location of the object in space. As long as the surfaces are parallel, the rays will not deflect.

A typical application of this method is determining the parallelism of Nd : YAG rods, to be used as the active medium in laser cavities [13]. The deviation from parallelism allowed for the end surfaces is 10 arcsec. Figure 7.10 shows deflectograms of a 4 mm diameter Nd : YAG rod at two positions achieved by 180° rotation. The arrows point at the same fringe to show its displacement during rotation. The accuracy reached at that case was ~ 3 arcsec.

Another application of beam deflection measurement is the direct determination of the birefringence of uniaxial crystals [14]. Uniaxial materials have two principal indices of refraction, n_o (ordinary) and n_e (extraordinary). The electric field of the ordinary propagating wave is everywhere normal to the optical axis and has a constant speed (c/n_o), whereas the electric field of the extraordinary propagating wave is parallel to the optical axis and its speed varies according to the orientation of the electric field with respect to the optical axis. The difference in the refractive indices $\Delta n = n_e - n_o$ is a measure of the birefringence of the material. To measure the birefringence by moire deflectometry the test sample is fabricated into a prism (see Fig. 7.11) so that the entering beam is not parallel to the optical axis. As a result the ordinary and extraordinary refractive beams exit the prism propagating at different angles ϕ_o and ϕ_e, respectively. Accordingly, two moire patterns are produced. The difference angle $\beta = \phi_o - \phi_e$ is found by adjusting the gratings' gap until an overlap between two fringes belonging to the two

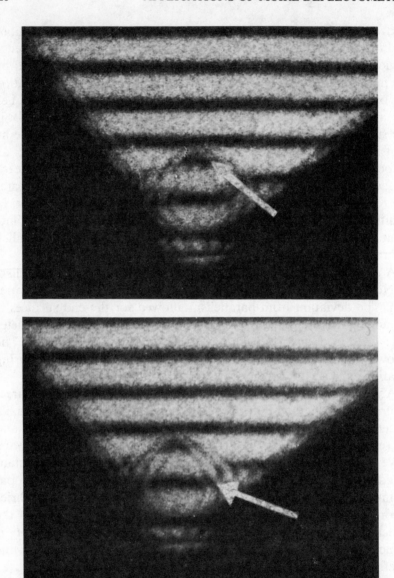

Figure 7.10. Deflectograms of a 4 mm diameter Nd : YAG rod at two extreme positions achieved by a 180° rotation. The fringe shift resulting from the wedge angle is indicated by arrows.

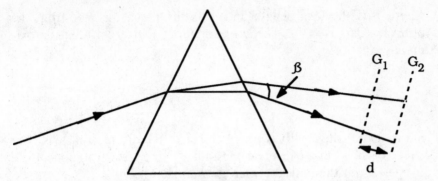

Figure 7.11. Setup for a moire deflectometric measurement of birefringence using a wedge of the birefringent material: G_1 and G_2 are the two gratings separated by a distance d.

patterns occurs (at a gap d'). The gap is varied again until the next overlap is attained (at d'') and the moire patterns are now shifted by exactly one fringe. β is given by

$$\beta = \frac{p}{d'' - d'}. \tag{7.13}$$

For small wedge angles, where $\alpha \ll \sin^{-1}(1/n_{o,e})$, the birefringence is given by

$$\Delta n = n_e - n_o \cong \frac{\beta}{2 \sin(\alpha/2)}. \tag{7.14}$$

7.3.2. Focal Length of Lenses and Mirrors

In the paraxial region, i.e., in the vicinity of the optical axis, the curvature of a spherical mirror or a lens can be approximated to that of a parabola. The ray deflection of such an element $\phi(y)$ is given by

$$\phi(y) \cong \tan^{-1}\left(\frac{y}{f}\right) \cong \frac{y}{f}, \tag{7.15}$$

where f is the focal length and y is the radial distance from the optical axis ($y \ll f$). The corresponding fringe shift is approximately

$$x' = \frac{yd}{\theta f},\qquad(7.16)$$

namely, the fringe shift is proportional to the distance off-axis. The effect of such an element on the moire pattern is a rotation of the fringes about the optical center at an angle α:

$$\alpha = \tan^{-1}\left(\frac{x'(y)}{y}\right) \cong \tan^{-1}\frac{d}{\theta f}.\qquad(7.17)$$

The focal length is easily derived from the rotation angle:

$$f \cong \frac{d}{\theta \tan \alpha}.\qquad(7.18)$$

At larger y values, where the paraxial approximation is no longer valid, an S-shaped bending of the moire fringes is observed (Fig. 7.12). From that fringe deformation the spherical aberrations can be evaluated. Moire deflectometry is particularly suited for measuring the focal length of very weak lenses or mirrors with a focal length of 100 m or even longer, for which other methods are not accurate enough or become impractical due to the very long optical benches required. In extreme cases where the fringe rotation caused by the lens is too small to be observed visually (below 10^{-2} rad) the moire of moire method can be applied. A reference deflectogram is superimposed on the deflectogram distorted by a lens and the pitch of the secondary fringes is inversely proportional to $\sin \alpha/2$.

The folded telescopic moire deflectometer provides a solution for measuring nonparaxial elements. A concave mirror can be measured by positioning it in place of the collimating object lens and the flat end mirror of the telescope (Fig. 7.3). The focal length is found by measuring the distance from the mirror vertex to the pinhole of the spatial filter, when the mirror position yields an optimally collimated beam on the deflectometer [15]. The accuracy in estimating the focal

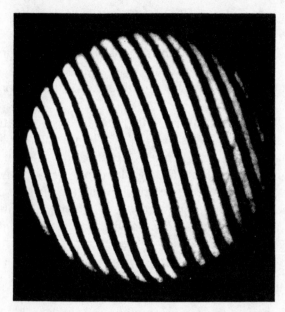

Figure 7.12. Moire deflectogram of a lens. The S-shaped fringes indicate spherical aberration.

length is limited by the fringe resolution, given by q, the minimum resolvable fraction of a fringe. The error δf is given by

$$\delta f \geq q \frac{p}{d} \frac{f^2}{a}, \qquad (7.19)$$

where a is the effective aperture. For example, the relative error for a mirror of length 350 mm is less than 10^{-4}. The focal length of a converging lens can be measured in the telescopic setup by varying the distance between the flat end mirror and the test lens until a collimated beam is produced on the deflectometer, i.e., the distance measured between the secondary vertex of the lens and the mirror face is now the back focal length f_b of the lens. Divergent lenses can be measured in a similar manner by replacing the flat end mirror by a concave mirror of known curvature. The distance between the mirror and the test lens is adjusted until the center of curvature of the mirror and the focal point of the lens coincide to result in a collimated beam on the deflectometer. To complete the picture we mention that convex mirrors can be tested in a similar way. The surface to be tested is positioned in place of the flat mirror and a converging lens is inserted in front of it at a distance where its focal point coincides with the center of curvature of the mirror to yield a collimated beam.

For low resolution measurements, one universal equation can be used for all lenses and lens systems [16]. The method applies the fact that the number of beat fringes obtained in the infinite fringe configuration when the beam slightly converges or diverges is a function of only the deflectometer parameters and the f number of the lens. Therefore, counting the fringes yields the f number. To find the number of fringes we assume that the lens causes the pitch of the first grating which is projected on the second grating to be changed by $(f - d)/f$. The beats pitch p' is $p(f - d)/d$. The aperture size a on the second grating is also changed by a factor of $(f - d)/f$. Therefore, the total number of fringes is ad/pf or d/pf-number. In this method the fringes can be counted with the lens simply inserted into the basic deflectometer setup.

The high accuracy in determination of the focal length of even very weak lenses enables the use of moire deflectometry in the study of physical phenomena that are known to slightly modify the focal length. In the next section, which deals with dynamic systems, we demonstrate the use of moire deflectometry in the study of the lens-like behavior of light absorbing media, known as the thermal lens effect. In this section we present some examples of static effects that were studied through the change in focal length that they induce.

7.3.2.1. Refractive Index of Fluids [11]. Moire deflectometry responds only to lateral gradients in the optical path of rays rather than to differences in the optical path itself. Therefore, in order to determine the refractive index of a homogeneous medium, a known gradient in the geometrical path has to be introduced. For example, when placed in a spherical glass vessel, a fluid whose refractive index differs from that of the ambient atmosphere performs as a lens. To measure the refractive index of fluids, gases, and liquids, a three-compartment cell, shown in Fig. 7.13, was designed. The inner space

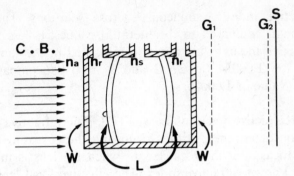

Figure 7.13. A lens-like three-compartment cell for measuring the refractive index of liquids and gases by moire deflectometry.

is confined by two identical lenses of diopter zero and the outer windows are optically flat. The central compartment is filled with a fluid of unknown refractive index n_s and the two outer compartments contain a reference fluid of known refractive index n_r. The refractive index of the glass is given by n_w and the radii of curvature of the inner and outer sides of the lenses are r_1 and r_2, respectively. Assuming a thin lens, the focal length of the lens formed is given by

$$f^{-1} = \frac{n_s - n_w}{r_1} + \frac{n_w - n_r}{r_2}. \qquad (7.20)$$

Since $r_1 = r_2 = r$ in a lens of diopter zero, we obtain an expression connecting n_s with the fringe rotation angle α:

$$n_s = n_r + \frac{r}{f} \cong n_r + \frac{r\theta \tan \alpha}{d}. \qquad (7.21)$$

The minimum detectable change in refractive index $\delta n_{\min}$ is set by the limit $\alpha_{\min} \geq p'/a$ (where a is the diameter of the lens),

$$\delta n_{\min} \cong \frac{\alpha_{\min} r\theta}{d} \geq \frac{rp}{ad}. \qquad (7.22)$$

The ultimate limit set by the uncertainty principle for $a = 30$ mm, $\lambda = 6 \times 10^{-4}$ mm is about 10^{-5}. The method was used to determine the refractive index of aqueous sucrose solutions. The error in measurement, compared to the literature values, is less than 10^{-4}. The refractive index of air was also measured in the same cell and found to be 1.00024 (at 27°C and 1 atm) as compared to the handbook value of 1.00026.

7.3.2.2. Refractive Index of Lenses [17].

The refractive index of a lens can be determined, without knowing the curvature of its surfaces, by measuring the focal length when the lens is immersed in liquids of known refractive index. The effective focal length f_i of a thin lens with unknown refractive index n_s, immersed in a liquid with refractive index n_i, is given by the lens makers formula

$$f_i^{-1} = \left(n_s - n_i\right)\left(\frac{1}{r_1} - \frac{1}{r_2}\right), \qquad (7.23)$$

where r_1 and r_2 are the radii of curvature of the lens faces. We denote the fringe rotation angle due to immersion of a lens in liquids with n_1 and n_2 by α_1 and α_2, respectively. Then n_s is given by

$$n_s = \frac{n_1 \tan \alpha_2 - n_2 \tan \alpha_1}{\tan \alpha_2 - \tan \alpha_1}. \qquad (7.24)$$

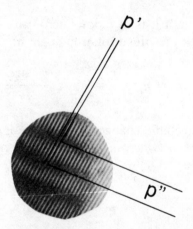

Figure 7.14.   Double exposure moire deflectogram of a benzene-filled cell with and without a lens. p' is the moire pitch and p'' is the moire of moire pitch.

Figure 7.14 is a double exposure moire deflectogram of a crown glass lens with a focal length of 100 cm in air, immersed in benzene. The reference deflectogram is of a benzene-filled cell. The refractive index found by successive immersion in benzene and toluene, $1.5210 \pm 5 \times 10^{-4}$, is in good agreement with the literature.

7.3.3. General Wave Front Distortion

In this category we include both transparent and reflective phase distorting objects, for which the interpretation of the moire pattern usually requires numerical analysis. The information derived refers to the optical quality of the test element, including properties such as spherical aberrations, local defects in manufacturing, nonuniformity of the refractive index, or deviation from surface flatness. Wave front distortion is the total peak-to-peak deformation of the emerging wave front from its unperturbed form.

Traditionally, these measurements are performed with an interferometer, in which the test object is compared against a reference of high optical quality (allowed surface error of about $\lambda/20$). The difference in the optical paths of the two wave fronts, seen as a deviation from the expected interference pattern, is caused by errors in the test element or tilt of one element relative to the other. Perfectly straight fringes indicate ideal quality while the fringe deviation from linearity is a measure of the wave front distortion. In the Twyman–Green interferometer, the spacing between two adjacent fringes is $\lambda/2$.

Moire deflectometry provides the lateral gradients of the optical path difference. Interferometric compatible information can be obtained by integration of the moire deflectometer data of ray deflection along the fringe direction. Quality control tests of both phase objects and flat surfaces have been accomplished by moire deflectometry.

7.3.3.1. Flatness Analysis.
The deviation from surface flatness has been studied extensively in order to reconstruct the surface's topographical map from its moire deflectogram. Figure 7.15 shows a moire deflectogram of a 5.25 in. aluminum hard disk substrate obtained with a 6 in. folded telescopic deflectometer with 7.5 ×

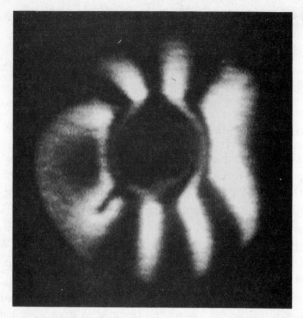

Figure 7.15. Infinite fringe moire deflectogram of a 5.25 in. diameter hard disk substrate.

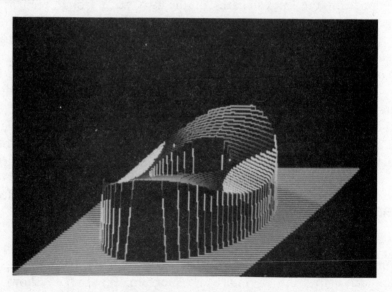

Figure 7.16. A numerically calculated contour map of the hard disk substrate of Fig. 7.15.

magnification. A contour map of partial derivatives (i.e., slopes) is obtained. In order to determine the surface topography, numerical integration of two deflectograms, providing the deflections along x and y axes, should be performed. Although the height contours obtained in interferometry may seem more straightforward to interpret, interferometry is highly sensitive to the object position and the fringe ordering is ambiguous. In order to obtain the phase information in interferometry, phase shift technology (see Chap. 2) is needed. This technique requires three photographs compared to the two finite fringe maps required in deflectometry. The moire deflectogram is recorded by a CCD camera and the digitized data is fed into an image processor for fringe analysis. Figure 7.16 is a numerically calculated contour map of the same disk. The increment between subsequent contours is 1 μm [18].

7.3.3.2. Aberrations. Wave front analysis by moire deflectometry is not restricted to flat objects, but can be applied even to highly

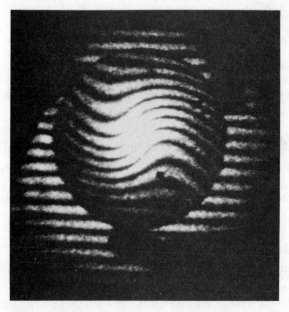

Figure 7.17. A moire deflectogram of a lens with a focal length of 29 cm. The S-bending of the fringes is due to spherical aberrations.

curved surfaces or strong lenses. The only condition is that the beam entering the deflectometer must be parallel and therefore, a correcting element of high optical quality is introduced to compensate for the curvature. Figure 7.17 is a moire deflectogram of a lens with 29 cm focal length, where only the S-bending of the fringes, due to spherical aberrations, is left after correcting for the lens effect. A detailed description of aberration analysis by moire deflectometry is given in the Appendix.

7.4. APPLICATIONS OF MOIRE DEFLECTOMETRY IN DYNAMIC SYSTEMS

In the applications presented in this section, ray deflection is caused by the formation of density gradients in the test media. Due to the approximately linear relationship between the density and the refractive index, the density, or temperature distribution, of the medium under test can be derived from the ray deflection map. Density gradients were mainly investigated in flow systems and in nonuniformly illuminated light-absorbing materials that show a thermal lens effect.

7.4.1. Flow Systems

In compressible media, ray deflections originate from gradients of the refractive index due to inhomogeneous density distribution. Such density gradients can be incurred by nonuniform temperature distribution in heat producing systems (for example flames) or by pressure gradients, as in acoustic phenomena like those occurring in shock tubes or in wind tunnels. If the chemical composition of the system investigated is known, the refractive index distribution can be translated to a density or temperature profile. The effect is demonstrated by the behavior of tap water in a glass [19]. Figure 7.18a shows a side view deflectogram of water in a closed rectangular glass container. Figure 7.18b shows the same object, with the container open, and we observe the bending of the fringes just below the water–air boundary layer. The bending builds up a few seconds after the lid is removed and its direction, indicating upward ray deflection, proves

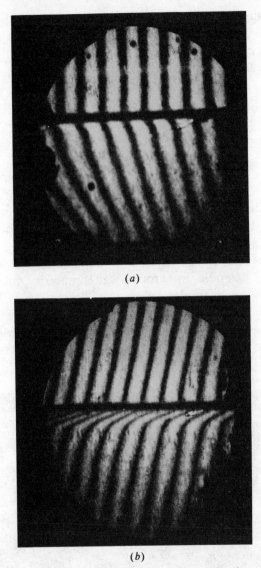

(a)

(b)

Figure 7.18. (a) A deflectogram of water in a closed rectangular container. (b) The same as (a) but with the cover removed. The bending of the fringes is due to evaporation cooling.

that the bending is caused by an increase in the density at the upper layer of the liquid phase toward the interface. This is interpreted as cooling of the upper water layer during evaporation, which is obviously enhanced upon opening the container. The linear shape of the bent fringes suggests a lens-like effect, according to which the temperature gradient in the vertical direction is related to the refractive index by

$$\frac{\partial T}{\partial y} = \frac{1}{n\gamma} \frac{\partial n}{\partial y}, \qquad (7.25)$$

where γ is the thermal expansion coefficient of water. The overall temperature decrease from the bulk to the interface was found to be 0.75°C.

Analysis of the density field is invaluable for the study of flow dynamics, for example, in aeronautical engineering and combustion diagnostics. Some examples follow.

7.4.1.1. Supersonic Wind Tunnels. The common means for visualization of flow patterns in shock tubes and wind tunnels is the

Figure 7.19. Schlieren photograph of a diamond-shaped 2-D airfoil in a supersonic wind tunnel at mach 1.98 and pressure 0.415 atm.

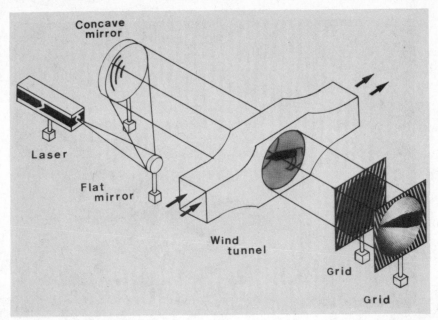

Figure 7.20.   Setup of the moire deflectometer for measurements in wind tunnel.

semiquantitative schlieren method, which distinguishes between rays deflected to opposite directions. The collimated beam that passes through the test chamber is refocused and a knife edge is placed so that it obstructs one half of the focal plane. All the rays that were deflected by positive gradients in the flow will focus on one side of the focal plane and all the rays that were deflected by negative gradients will focus on the opposite side. Therefore, blocking one-half of the focal plane will darken all the areas from which rays are deflected in that direction. Due to diffraction effects and shadowing, the larger slopes in the unblocked direction appear brighter than the smaller ones and semiquantitative shadowing appears. Figure 7.19 is a schlieren photograph of a diamond-shaped 2-D airfoil in a supersonic tunnel at mach 1.98 and pressure of 0.415 atm [20].

Moire deflectometry may be regarded as a "quantitative schlieren" method that supplies not only the general direction of rays, but also the exact angle of ray deflection. In practice, a schlieren setup can be easily transformed into a moire deflectometer by replacing the refocusing optics needed for the spatial filtering by a pair of Ronchi

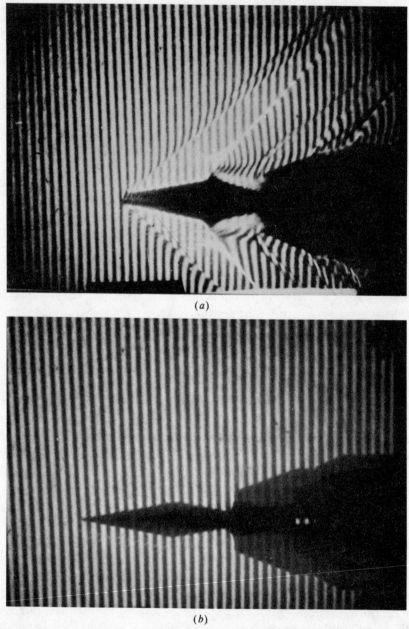

(a)

(b)

Figure 7.21. (a) Moire deflectogram of the same object as in Fig. 7.19 with the same flow conditions. (b) The fringe pattern observed in the absence of flow.

rulings as shown in Fig. 7.20. Figure 7.21*a* is a moire deflectogram of the same object shown in Fig. 7.19, subjected to the same flow conditions [20]. The fringe pattern serves as a built-in ruler. Moire deflectometry detects not only drastic density changes as in schlieren, but also the infrastructure of each zone in the picture. For comparison, Fig. 7.21*b* depicts the fringe pattern observed in the absence of flow.

To demonstrate the usefulness of the method in density field analysis, the density gradients at numerous points along the expansion fan of the wake have been calculated for an ideal flow pattern around the same diamond-shaped object at the same angle of attack. The comparison between the measured data derived from the fringe shift with respect to the position at rest and the calculated density gradients, shows a good agreement with the ideal model except for a small turbulent zone close to the sample holder. Thus, measurement of the density gradients by moire deflectometry can save tedious calculation in the case of complicated flow patterns.

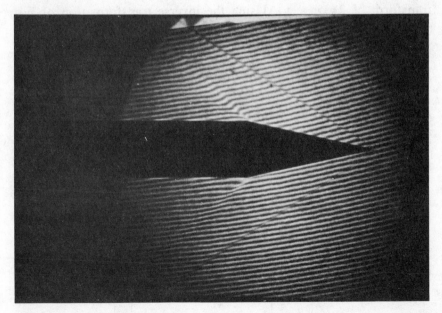

Figure 7.22. Moire deflectogram obtained with a circular cone held at zero angle of attack in a supersonic wind tunnel at mach 1.98 and pressure 3.15 atm.

The analysis of 3-D patterns seems to be more complicated, particularly in the case of asymmetric density distributions. However, when the phase object exhibits some symmetry, the computation can be quite simplified. For example, we show the treatment of a flow field with axial symmetry. The sample studied is a circular cone held at zero angle of attack in a supersonic wind tunnel [21]. A deflectogram of the flow pattern past the cone, at mach 1.98 and pressure of 3.15 atm, is presented in Fig. 7.22. The flow is compressed by an oblique conical shock wave and remains constant on cones having the same vertex, except for the boundary zone. For an object with axial symmetry $n(x, y, z) = n(r, y)$ where $r = (x^2 + z^2)^{1/2}$, ϕ can be rewritten in cylindrical coordinates (r, ξ, y) as

$$\phi(x, y) = \frac{2x}{n_f} \int_x^{r_s(y)} \frac{\partial n(r, y)}{\partial r} (r^2 - x^2)^{-1/2} dr, \qquad (7.26)$$

where $r_s(y)$ is the radius of the shock wave. Equation (7.26) is called the Abel transformation and therefore $\phi(x, y)$ is the inverse Abel transform of $2\pi x n(r, y)$. Using the inverse transformation of Eq. (7.26) n is found to be

$$n(r, y) = n_f \left[1 - \frac{1}{\pi} \int_r^{r_s(y)} \phi(x, y)(x^2 - r^2)^{-1/2} dx \right]. \qquad (7.27)$$

Assuming that n is constant on cones with a common vertex at the apex of the object, $n(r, y)$ can be substituted by $n(a)$, where $a = r/y$. Accordingly, $r_s(y)$ is replaced by $a_s = r_s/y$ and x by $b = x/y$ and a simpler expression for n is formed:

$$n(a) = n_f \left[1 - \frac{1}{\pi} \int_a^{a_s} \phi(b)(b^2 - a^2)^{-1/2} db \right]. \qquad (7.28)$$

Since $\phi(b)$ is proportional to b (as the straight fringes indicate), the integration yields

$$n(a) = n_f \left[1 - \frac{1}{\pi} \frac{k\theta}{d} (a_s^2 - a^2)^{-1/2} \right], \qquad (7.29)$$

where k is a constant satisfying $y'(x, y) = -kb$. Figure 7.23 demonstrates the congruence between the measured density distribution and the theoretical profile. As manifest in this example, the data furnished by moire deflectometry, namely, gradients of the refractive index rather than the refractive index itself (as in interferometry), are exactly the kind of information directly usable by the Abel transformation. For the specific application of deriving a 3-D density gradient distribution from a given 2-D distribution, moire deflectometry is particularly suitable. This is true not only in axisymmetrical objects, but also for totally asymmetric objects.

To find the density distribution in the lack of any symmetry, a series of moire deflectograms produced at different projections of the object with a total viewing angle of 180° is required. The mathemati-

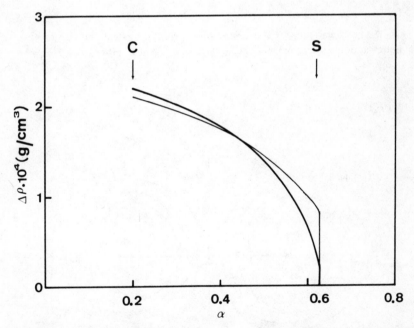

Figure 7.23. Variation of the density $\Delta\rho$ vs. α, the tangent of the coordinate angle α. α_c and α_s denote the cone angles of the object and the shock wave, respectively. The thin line and the thick line represent the theoretical and experimental results, respectively.

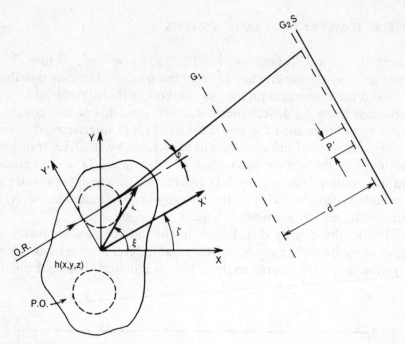

Figure 7.24. Schematics of the setup and coordinate system for 3-D analysis of asymmetric objects. From J. Stricker *Appl. Optics* **23** 3657 (1984).

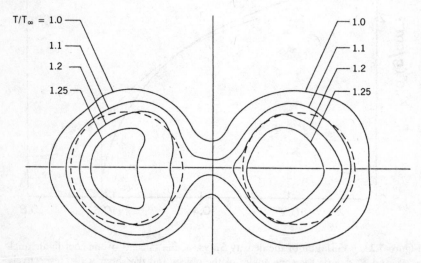

Figure 7.25. Temperature contour map around two vertical heated cylinders, 6 cm apart, obtained from numerical integration of the Radon transform at 10 viewing angles. T_∞ is the ambient temperature. From J. Stricker *Appl. Optics* **23** 3657 (1984).

cal means for inverting the integrated deflection angle into a refractive index distribution is the Radon transform [22]. The refractive index distribution in a particular plane of constant z is given in cylindrical coordinates by

$$n(r, \xi, y) - n_\infty = \frac{1}{2\pi^2} \int_{-\pi/2}^{\pi/2} d\zeta \int_{-\infty}^{\infty} \frac{\phi(x, y, z)\, dx}{r \sin(\xi - \zeta) - x}, \quad (7.30)$$

where ζ is 90° minus the viewing angle, as shown in the schematics of the experimental setup (Fig. 7.24). For demonstration purposes a flow field with a plane of symmetry, generated by two vertical heated cylinders 6 cm apart, was investigated. Due to the plane of symmetry a 90° viewing angle is sufficient. The density distribution, obtained by numerical integration of the Radon transform from 10 viewing angles (at 10° intervals) at a plane $y = 0.7$ mm about the top of the cylinders, was further converted to temperature distribution ($T = k\rho^{-1}$, where k is a constant and ρ the air density). Figure 7.25 is a temperature contour map derived from the temperature profiles at each viewing angle.

7.4.1.2. Flame Temperature Distribution. As seen in the previous section, the temperature distribution of a medium with known chemical composition can be evaluated from the refractive index profile, in the same way as the density. Mapping of the refractive index of flames is similar to wind tunnel measurements except for the need in some cases to filter the noise caused by the flame's emission. The conversion of refractive index data into temperature is rather cumbersome because it requires detailed knowledge of the chemical composition at each point, an almost impossible mission. Nevertheless, in many cases the composition can be reasonably assumed or its effect is secondary (for example, when the constituents resemble in their molar refractivity). Since there exists no other straightforward means of simultaneous mapping of flame temperature, moire deflectometry offers, despite the slight ambiguity in the interpretation, a considerable solution to real time temperature mapping of flames. The method was demonstrated on flames with various temperature profiles, using different techniques of fringe analysis.

7.4.1.2.1. Premixed Hydrogen–Oxygen Flame [23]. The temperature profile of a flat flame of premixed hydrogen and oxygen was
investigated using a circular burner with 300 holes. The gas mixture
was enriched in oxygen to ensure that the combustion was practically complete immediately above the burner's exit, and the main
constituent of the flame was water. The refractive index profile of the
axisymmetric flame is derived by applying the Abel transformation
to the deflection angle. The deflection angle was found by measuring
the fringe shift in a modified way. Instead of direct measurement of
the fringe deviation, the transmittance of the deflectometer is scanned
along a horizontal slice of the upright flame (the undistorted fringes
are also horizontal). This is accomplished by using a video camera
equipped with a line selector. To convert the line densitogram to
deflection angle distribution, a triangular fringe profile is assumed so
that within one-half of the grating's period the fringe shift is linearly
proportional to the transmittance. An oscilloscope trace of the line
densitogram at a given cross section at some distance downstream is
given in Fig. 7.26. The relation is reversed at the second half and

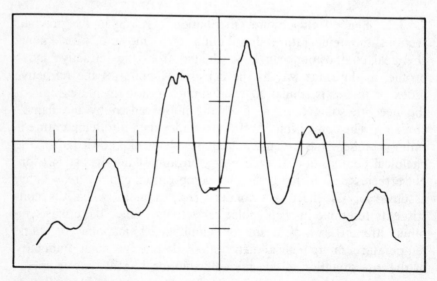

Figure 7.26. Densitogram of the moire pattern across a H_2–O_2 flame some
distance downstream.

repeated periodically. The relation between the deflection angle and the fringe shift, expressed as a phase shift ψ (i.e., a fringe shift of a complete pitch is analogous to a phase shift of 2π), is

$$\phi = \frac{p\psi}{2\pi d \cos(\theta/2)}. \tag{7.31}$$

The refractive index of a gas mixture can be approximated by a sum of the partial contributions of the components

$$n = 1 + \sum_i \kappa_i N_i, \tag{7.32}$$

where κ_i is the molar refractivity of the ith species and N_i is the molar concentration. Assuming ideal gas behavior, the temperature is derived using the gas law

$$T = \frac{P_t}{R(n-1)} \kappa_a, \tag{7.33}$$

where P_t is the total gas pressure, R is the gas constant, and

$$\kappa_a = \frac{\sum \kappa_i N_i}{\sum N_i}$$

is the average molar refractivity of a gas mixture. Since the molar refractivities of the main components in the flame are similar to within 10%, the calculation was carried out assuming that water is the only constituent. The error in estimating the temperature from the refractive index increases as the square of the temperature

$$\delta T = \frac{R}{P_t} \frac{1}{\kappa_a} T^2 \delta n. \tag{7.34}$$

This is demonstrated by the dashed area in Fig. 7.27, which is the radial temperature profile of the H_2–O_2 flame. The dot on the flame axis is a point measured by a thermocouple.

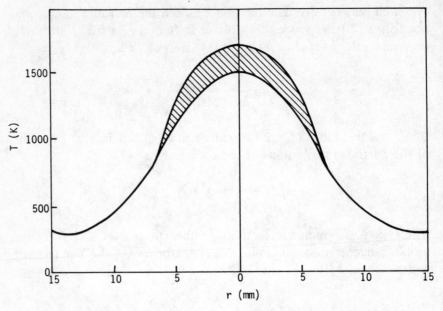

Figure 7.27. Temperature profile of the H_2-O_2 flame of Fig. 7.26.

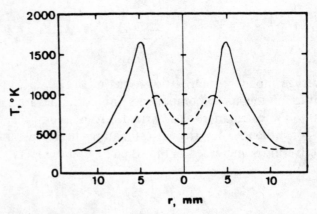

Figure 7.28. Radial temperature profile of a CH_4-air flame. The solid line curve was obtained at a distance downstream of 8 mm and the dashed line curve at a distance of 38 mm.

7.4.1.2.2. Methane–Air Flame [24]. A premixed methane–air flame shows a different temperature profile than that demonstrated by hydrogen. Again, a circular burner was used and the flame is assumed to be axisymmetric. Since the combustion of methane in air is much slower than that of hydrogen in pure oxygen and the rate is diffusion-controlled, we expect that the reaction proceeds faster in the boundary zone between the unburnt methane emerging from the burner and the ambient atmosphere. Figure 7.28 shows radial profiles of the measured temperature at two points downstream. As predicted, the temperature reaches a maximum at the envelope of the methane stream and decreases to a minimum at the center, which probably contains only air cooled by the sudden decompression. As the flame propagates downstream, the temperature falls and the distribution flattens.

7.4.2. Thermal Lensing

When a laser beam with a nonuniform intensity distribution is fired into an absorbing medium, lens-like behavior is observed. A Gaussian-shaped beam, for example, induces a spherical diverging lens. The thermal lens is generated by the heat-releasing nonradiative relaxation processes, subsequent to the absorption of light, which leads to a buildup of temperature gradients and a nonuniform refractive index distribution [25]. When a sample is illuminated by a continuous wave source, the thermal lens effect reaches a stationary state after a finite rise time. On the other hand, when the source is modulated or pulsed, the thermal lens reaches a peak and decays at rates that are characteristic of the molecular relaxation rates and heat transfer processes in the irradiated medium, respectively. The power of the thermal lens, i.e., the reciprocal focal length, depends on the amount of energy absorbed, which is proportional to the absorbance of the species at the wavelength of the light source, as well as to the laser power and absorbing species concentration. The temporal history of the thermal lens is mainly investigated by measuring the radiant flux density at the center of a second, nonabsorbed, laser beam used for the analysis, applying a pinhole. The beam width is modified on passing the thermal lensing medium. If one assumes an ideal thin lens behavior, the focal length is derived from the change in radiant flux density [25].

The thermal lens effect is used mainly to monitor weak absorptions with an extinction coefficient ε of $10^{-4}\,\mathrm{cm}^{-1}$ and below, and to study the absorption spectrum and relaxation mechanisms of weak absorbers [26]. Other means for thermal lens detection, such as interferometry, were suggested [27] but did not become widespread. In contrast to the pinhole technique, moire deflectometry provides a map of the density gradients over the entire illuminated area without previous assumption about the geometry of the effect. The application of moire deflectometry to thermal lens detection has been demonstrated in study of rise time and decay rates both in gas and solid state samples.

7.4.2.1. CO$_2$ Laser Induced Thermal Lens in SF$_6$ [28].

The strong absorption of SF$_6$ gas at $943\ \mathrm{cm}^{-1}$ of $\varepsilon = 5.6 \times 10^2\,\mathrm{cm}^{-1}\,\mathrm{atm}^{-1}$ (the ν_3 transition band) coincides with the main emission frequency of the nontunable CO$_2$ laser and gives rise to a strong thermal lens

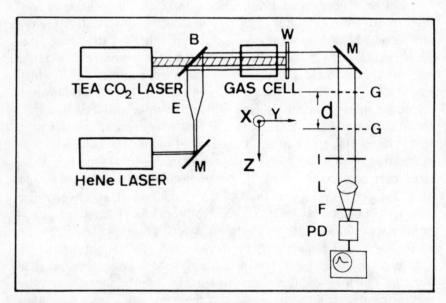

Figure 7.29. Moire deflectometer setup for thermal lens analysis in gases. M is a mirror, E is a beam expander, B is a beam mixer, W is an infrared-radiation-blocking window, G is a Ronchi ruling, I is an iris, L is a lens, F is an interference filter, and P.D. is a photodiode.

effect. The experimental setup for moire deflectometer analysis of the thermal lens is given in Fig. 7.29. The exciting beam of the pulsed CO_2 laser (marked by the dashed zone) and the analyzing broad beam of a He–Ne laser propagate collinearly into the gas cell which is filled with a mixture of SF_6 and He. At the cell exit the IR beam is blocked while the He–Ne beam proceeds into the deflectometer.

The deflectometer is set at infinite fringe configuration so that the transmittance at each point is proportional to the deflection angle, assuming a triangular fringe profile. To record the short lived effect (the exciting pulse duration is $\sim 10^{-5}$ s) fast-response detection is required. A single photodiode detector was used so that mapping of the lens area requires 2-D scanning with an aperture. A typical oscilloscope trace of the transmittance at a distance 3 mm off-axis is shown in Fig. 7.30. We define $\phi = 0$ for an initially dark fringe pattern (with transmitted intensity I_{min}). Therefore, a rise in the transmitted intensity to a value I (within one-half of a fringe) means a deflection angle ϕ, where

$$\phi = \frac{p}{2d} \frac{I - I_{min}}{I_{max} - I_{min}}, \qquad \phi < \frac{p}{2d}, \qquad (7.35)$$

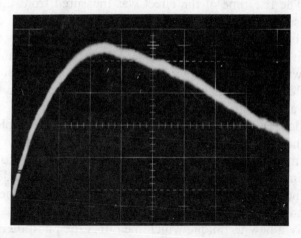

Figure 7.30. A typical oscilloscope trace of the thermal lens signal obtained by the moire deflectometer when SF_6 is irradiated by a CO_2 laser pulse. Off-axis distance is 3 mm and time base is 0.1 ms/div.

and I_{max} is the maximum intensity at a phase shift of π. If, however, the initial phase shift is $\pi/2$, which gives rise to an intensity $I_0 = \frac{1}{2}(I_{max} + I_{min})$, then the direction of ray deflection is given by the sign of the change in intensity.

If the thermal lens is scanned with a circular aperture of radius r along the y direction, then the transmitted energy S collected at a distance R from the center of the beam is given by

$$S = \int_{R-r}^{R+r} dy \int_{-[r^2-(y-R)^2]^{1/2}}^{[r^2-(y-R)^2]^{1/2}} (I - I_0)\, dx = \frac{\pi R r^2 d}{pf}(I_{max} - I_{min}).$$
$$(7.36)$$

In this calculation $\phi(y)$ is replaced by y/f, the paraxial approximation. The result of a scan over the y axis (i.e., a cross-section intensity profile), performed with a 1.2 mm wide aperture, is shown in Fig. 7.31a. The sign of $I - I_0$ is inverted on crossing the center, as expected. Integration of $I - I_0$ over the beam radius yields the profile of the refractive index, and hence, the spatial distribution of the reciprocal temperature. In Fig. 7.31b the solid line represents the numerically integrated S, while the dots show the significantly narrower profile of the pump laser monitored by a photon drag detector. The rise time of the effect was measured from the buildup rate as observed on the oscilloscope. Since the temporal behavior is found to be exponential, the characteristic rise time is merely the time required for the signal to reach $1/e$ of the maximum. Figure 7.31c shows a rise time dependence on the off-axis distance. Thus, moire deflectometer analysis of a thermal lens provides both the deflection angle distribution and the 2-D time evolution of the effect.

Moire deflectometry is superior to detection by a pinhole because of the higher signal to noise ratio (SNR), which in moire deflectometry is defined as the ratio of the transmitted intensity I to I_{min}. The ultimate limit is I_{max}/I_{min}. In the pinhole method SNR is determined by z/f, where z is the distance between the sample and the detector, which cannot be increased without limit.

Recently a new technique which measures the photothermal beam deflection has been applied to thermal lens analysis [29]. In this technique the deflection of a narrow probe beam, which intersects

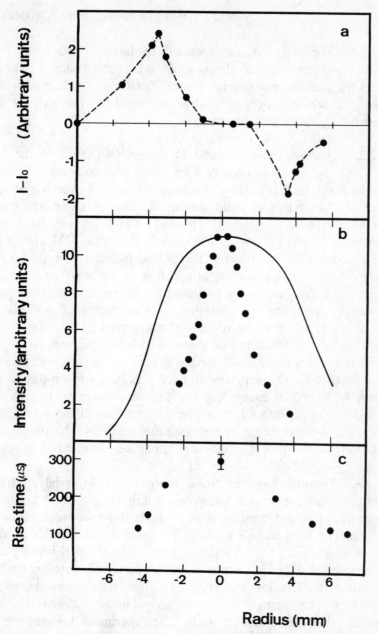

Figure 7.31. (*a*) Thermal lens signal vs. off-axis distance of the sample in Fig. 7.30. (*b*) The solid line curve indicates the thermal lens profile obtained by numerical integration of the signal of Fig. 7.31*a* and the dots indicate the CO_2 laser beam profile. (*c*) Rise time of the thermal lens signal vs. radial distance.

the pump beam inside the sampled medium, is measured by a position sensitive optical detector. If the probe beam is scanned across the sample, one obtains a profile of the photothermal deflection signal which is proportional to the ray deflection, exactly as is measured by moire deflectometry.

7.4.2.2. Thermal Lens in Solid Hydrogen [30]. Due to the high sensitivity, moire deflectometry is an ideal method for measuring the thermal lens effect caused by the weak vibrational overtone absorptions of, for example, solid hydrogen. The overtone spectra of normal and parahydrogen (n-H^2 and p-H^2) crystals have been extensively studied using photoacoustic detection [31] and FTIR spectroscopy [32]. Monitoring the pulsed induced thermal lens by moire deflectometry provides, in addition to the overtone spectrum, the nonradiative relaxation rates and the rates of heat transfer processes subsequent to excitation. The excitation of p-H_2 to the first and second overtone states was accomplished by a light pulse produced by a frequency-doubled Nd : YAG pumped dye laser whose frequency is Stokes-shifted to the IR region by a high pressure H_2 Raman cell. The experimental setup is basically similar to that shown in Fig. 7.29 except for the different pumping system. The photodiode is fed into a fast-response current amplifier and digitized by a fast transient recorder for computer analysis. The preparation of a p-H_2 crystal free of cracks and impurities is described in ref. 33.

7.4.2.3. Thermal Lens in Solid State Laser Materials. When a lasing medium, such as a laser rod, is uniformly excited by optical pumping, the heat generated is dissipated by convection to the surface. The temperature gradient produces a radial index gradient dn/dr which gives rise to a thermal lens. The thermal lensing effect is detrimental to the performance of the laser, and therefore study of the effect is of major importance in solid state lasers design and operation. The thermal lensing in several laser materials has been studied using moire deflectometry. The experimental arrangement is shown in Fig. 7.32 [34]. The laser rod is mounted in a double elliptical pump chamber where it is uniformly pumped by two flash lamps. The rod surface is kept at a fixed temperature. The analyzing beam of an He–Ne laser is distorted on passing along the rod and

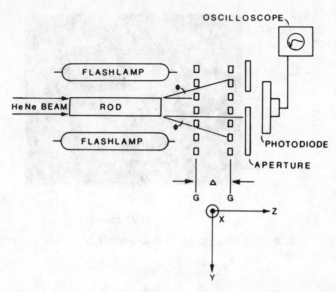

Figure 7.32. Moire deflectometer setup for thermal lens analysis in laser rods.

the deflectometer's transmittance is collected by a photodiode. To allow discrimination between a convergent and a divergent thermal lens a semicircular aperture with a radius r, which allows the passage of upward directed rays only, is positioned on-axis before entering the detector. The transmitted energy S is

$$S = \int_0^r dy(I - I_0) \int_{-(r^2-y^2)^{1/2}}^{(r^2-y^2)^{1/2}} dx = \frac{4r^3d}{3pf}(I_{max} - I_{min}). \quad (7.37)$$

Figure 7.33 is an oscilloscope trace of the thermal lens signal of an alexandrite rod, in which the intensity is already translated to units of lens power, i.e., diopters. The temporal history of the thermal lens comprises two phases. At the short first stage the thermal lens is strongly divergent. It decays to zero and builds up again on a longer time scale, as a convergent lens.

Since the dominant source of thermal lensing in laser rods is the temperature dependence of the refractive index, attempts are focused on finding athermal lasing materials, i.e., crystals in which $dn/dT =$

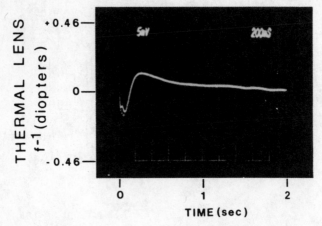

Figure 7.33. Oscilloscope trace of the thermal lens signal (in lens power units) of an alexandrite rod.

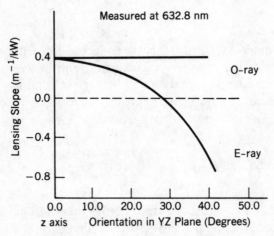

Figure 7.34. Thermal lensing in Nd : BEL laser rods for various crystal orientations.

0. A possible candidate is Nd : BEL (neodymium-doped lanthanum beryllate crystal) [35]. The monoclinic Nd : BEL crystal is optically biaxial, and it was found that dn/dT is strongly dependent on the orientation of the extraordinary ray path with respect to the principal optical axes of the crystal. A series of Nd : BEL rods of varying crystallographic orientation were tested by moire deflectometry, and the dependence of the lens power per kilowatt pump laser on the crystal orientation is shown in Fig. 7.34. At a certain orientation the thermal lens vanishes and this athermal rod can provide a better performing laser.

7.5. APPLICATIONS OF MOIRE DEFLECTOMETRY TO MTF ANALYSIS

7.5.1. Experimental Techniques

The local MTF [36] (i.e., the fringe modulation at a given point) is mapped by scanning the transmittance across a finite fringe pattern with a photodiode. To predetermine the diffraction-limited divergence, a pupil of known aperture is introduced before the gratings. The MTF of the test object is found by normalizing the modulation according to the reference aperture. The field of view of the photodiode is limited by a narrow slit whose width is less than one-tenth of the fringe period. Its length is much larger than the pitch of the grating over which the signal is averaged and smaller than the local fluctuations in the fringe position. Another way of averaging over a grating's pitch, which is useful if the fringes are not straight, is to move the two gratings in unison in a direction perpendicular to the grooves at a constant speed [37]. This eliminates the high frequency component of the moire pattern and compensates for local defects of the gratings without losing the spatial resolution (see Chap. 2). To avoid diffraction effects of the finite pupil one should not measure closer than $\lambda d/2p$ to the edge.

In most cases found in literature the quality of an object is specified by its overall, rather than local, MTF. The overall MTF is usually found from the point spread function of the object by applying Fourier transformation or by autocorrelating the pupil

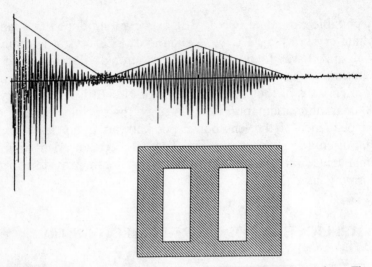

Figure 7.35. Experimental MTF curve of a double slit aperture vs. shear. The solid envelope is the theoretical MTF, equivalent to the correlation function.

function by applying lateral shearing interferometry (i.e., notice the way in which Melles Griot specifies the MTF of lenses [38]).

To obtain the overall MTF of a test object by moire deflectometry, the deflectometer is set to infinite fringe mode and the overall transmittance is collected and focused on the detector. The reduced spatial frequency is tuned by changing the gap between the gratings, and the MTF curve is given by drawing the overall transmittance vs. the gap d [36]. Figure 7.35 shows the integrated light intensity obtained with a double slit aperture vs. the spatial frequency ($p = 0.08$ mm, d varied between 0 and 100 cm). The solid envelope is the calculated autocorrelation function, which is the theoretical MTF for this aperture. In the applications presented, however, the local MTF is measured.

It is worth noting that the definition of the MTF in Chap. 6, which is used here, takes only the first diffraction order into consideration. Higher diffraction orders may introduce an error up to $\sim 10\%$. This error can be reduced by using a telescopic setup. A reader interested in a more comprehensive discussion on the ray deflection approach to MTF analysis is referred to ref. 36.

7.5.2. Applications of MTF Analysis in Static Systems

7.5.2.1. Quality Analysis of Laser Beams [39].

The number of transverse modes of radiation n serves as a quantitative criterion for beam quality, specifically, how well can a beam be focused. As was shown in Eq. (6.48), n is related to the modulation M by

$$n = \left[\frac{(1 - M)pa}{\lambda d} \right]^2, \tag{7.38}$$

where a is the diameter of the beam. Beam quality analysis was demonstrated on a copper vapor laser built at the Nuclear Research Center–Negev. The laser operates at two optical configurations: a stable resonator mode consisting of two flat end-mirrors and an unstable resonator mode comprising concave and convex mirrors in a confocal position. In the second configuration the output coupler is

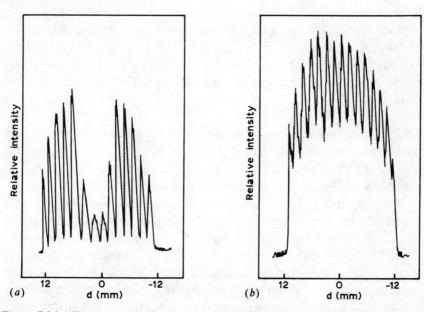

Figure 7.36.   Densitograms of moire patterns produced by beams of a copper vapor laser. (*a*) Unstable resonator. (*b*) Stable resonator.

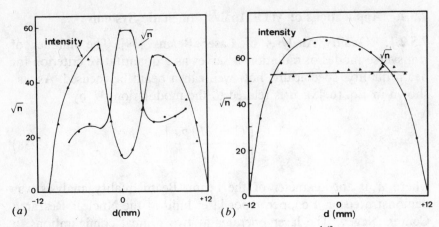

Figure 7.37. Radial profiles of the relative intensity and $n^{1/2}$ of a copper vapor laser beam. (a) Unstable resonator. (b) Stable resonator.

an annular flat mirror. Densitograms of the moire pattern of the laser beam at the two configurations are shown in Fig. 7.36, and the respective radial profiles of $n^{1/2}$ together with the relative intensity profile are provided in Fig. 7.37. The characteristic mode number $\langle n \rangle$, defined by averaging n over all the fringes, has a value of about 3000 in the stable resonator and around 800 for the unstable resonator.

7.5.2.2. Surface Roughness Analysis [40]. Surface quality is usually measured mechanically by a profilometer, a scanning probe for tracing the surface structure. The average roughness R_a (defined as the RMS of the height variation) is evaluated from the fluctuations in the height of the stylus along the track. It was suggested to determine the surface quality from the fringe contrast of the moire pattern produced by a beam reflected from the surface. The contrast degradation indicates the reduction of beam quality or rather the increase in the number of transverse modes of the beam upon reflection from a given surface. Because moire deflectometry is sensitive to ray deflections in one direction only, it can discriminate between the axes of polished surfaces. Thus, $n_{\parallel}$ and $n_{\perp}$, the

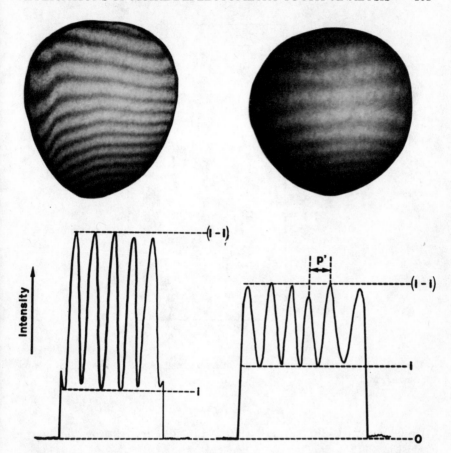

Figure 7.38. Moire deflectograms and respective intensity profiles of two stainless-steel polished plates.

number of transverse modes measured along the lay and perpendicular to it, respectively, can be determined independently by rotating the surface under test by 90°.

Moire deflectograms of two polished stainless-steel plates differing in quality, are shown in Fig. 7.38, together with the respective densitograms. Figure 7.39 presents the correspondence between the average roughness R_a measured by a profilometer with a 5 μm diameter stylus and the averaged value of $n^{1/2}$ derived from the

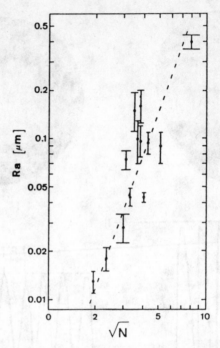

Figure 7.39. Correlation between the average roughness R_a measured by a profilometer, and $n^{1/2}$.

fringe contrast, for several plates. Linear curve fitting yields the relation

$$n \propto R_a^{0.7}. \tag{7.39}$$

7.5.2.3. Laser Rod Analysis [41]. The performance of a solid state laser depends on the chemical composition (i.e., dopant concentration) and optical properties of the crystal rod which serves as the active medium. For economical reasons it is desirable to specify the optical quality of the crystal boule in order to locate the best site out of which the laser rod will be cut. The conventional criterion is the degree of wave front distortion, but this alone does not guarantee high laser performance; thus MTF analysis has been suggested. Contrast reduction of the moire pattern of a He–Ne laser beam transmitted through alexandrite rods is demonstrated in Fig. 7.40a.

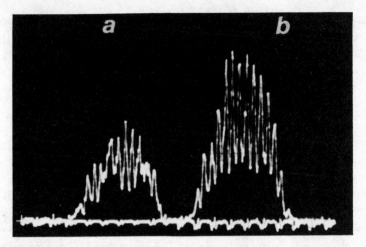

Figure 7.40. Moire deflectograms of (*a*) an alexandrite rod through an aperture and (*b*) aperture only.

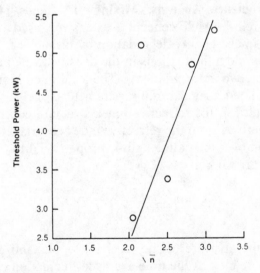

Figure 7.41. Threshold power as a function of $n^{1/2}$ for several 5 mm diameter alexandrite laser rods.

Figure 7.40*b* is a moire pattern densitogram of a reference aperture of the same diameter. The increase in the number of spatial modes *n*, on passing through the rods, is given by

$$\sqrt{n} = \frac{1 - M}{1 - M_{\text{ref}}}, \qquad (7.40)$$

where M is the modulation measured with the rod and M_{ref} is the modulation of the aperture alone. Since the performance of a cw alexandrite laser is mainly determined by the threshold power (which affects both the efficiency and the persistence of the rod), the correlation between $n^{1/2}$ and the threshold power was investigated. Figure 7.41 demonstrates the almost linear relationship between $n^{1/2}$, calculated for several rods, and the threshold power of cw lasers constructed from these rods.

7.5.3. Applications of MTF Analysis in Dynamic Systems

7.5.3.1. Turbulence Analysis: Mixing of Liquids [42]. In atmospheric research the MTF concept is extensively used as a qualitative criterion for image quality degradation by turbulence. The turbulent flow causes random fluctuations in the density of the medium, which result in refractive index gradients. The local microscopic deviations can be translated to measurable parameters, such as beam divergence, through C_N^2, the refractive index structure constant [43]. This constant is derived from the refractive index covariance function and its integration over the path length is proportional to the variance of the angle of arrival σ_α^2,

$$\sigma_\alpha^2 = 2.92 a^{-1/3} \int_0^L C_N^2(z) \, dz, \qquad (7.41)$$

where a is the aperture of the receiving optics and L is the path length. σ_α^2 is defined as the time-averaged fluctuation of the position of an image of a given object, a bar-test chart for example. By definition, σ_α^2 is equivalent to the solid angle divergence $\delta\Omega$. Therefore, C_N^2 can be calculated from the fringe contrast of the moire pattern.

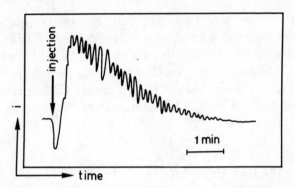

Figure 7.42. A typical recording of the transmittance of a dark moire fringe during the mixing of glycerol (95% in water) with pure water. Stirring speed is 2100 rpm.

The principle is demonstrated in a study of the mixing rate of two miscible liquids, water and aqueous glycerol solution, where C_N^2 is a criterion for the goodness of mixing. The goodness of mixing is determined by the homogeneity of the mixture, which is manifest by the fringe contrast. A typical recording of a mixing process is shown in Fig. 7.42. The transmittance of an initially dark fringe is monitored vs. time. The instantaneous drop of the water transmittance upon addition of the solute (95% glycerol) results from total light scattering. The following rise to a peak (which equals $[I_{min} + I_{max}]/2$ where I is the transmitted intensity [42]) indicates complete reduction of the fringe contrast. The exponential decrease to a fluctuation-free plateau at the same level as that of the pure solvent shows that the initial contrast has been regained.

The mixing rate τ^{-1}, where τ is the time required to fall to $1/e$ of the peak transmittance, was found to be proportional to the mechanical stirring speed, which is theoretically expected, and proves the reliability of this method.

7.5.3.2. Colloid Concentration and Particle Size
Determination—MTF Propagator [44]. The reduction of fringe contrast by a scattering medium, such as colloidal suspensions, can serve as a measure of the cross section for particle scattering. The transmittance of the moire pattern includes contributions of the unscattered intensity attenuated by absorption according to Beer's law and

of that part of the scattered light that reaches the second grating. While the unscattered intensity projects a diffraction-limited shadow of the front grating, the contribution of the diffused light is assumed to be totally blurred. Denoting the total cross section for absorption and scattering by σ_t and the fraction of diffused light that falls on G_2 (the second grating) by u, we obtain an expression for the MTF,

$$1/M - 1 = u(e^{\sigma_t \rho L} - 1), \tag{7.42}$$

where ρ is the number density of scattering particles and L is the cell length. The colloid studied was polybead polystyrene monodisperse latex of diameter 0.97 μm. Figure 7.43 shows the variation of $(1/M - 1)$ vs. colloid concentration. The solid line is the best fit calculation obtained with $u = 0.022$ and $\sigma_t = 7.3 \times 10^{-9}$ cm^2. The calculated cross section is practically identical to the given geometrical cross section of the particles. This means that each photon

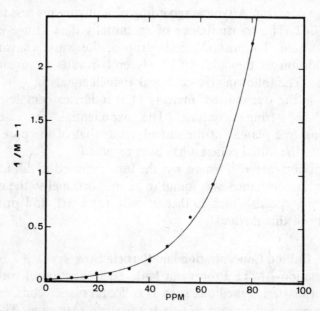

Figure 7.43. Plot of $(1/M - 1)$ vs. colloid concentration. The solid line is the best-fit calculation of the right-hand side of Eq. (7.42) and the dots represent experimental results.

colliding with a particle is either scattered or absorbed, as assumed in developing Eq. (7.42).

7.6. MISCELLANEOUS APPLICATIONS OF MOIRE DEFLECTOMETRY

Moire deflectometry was proven a useful tool in detection of small tilt angles in a variety of applications that are not directly connected with testing of optical properties of phase objects or specular surfaces. Some of those applications will be presented now.

7.6.1. Level Based on Moire Deflectometry [45]

The proposed optical level is shown in Fig. 7.44. A collimated beam is obliquely incident on the level, which consists of a flat mirror and an open container filled with liquid. If the measured surface is absolutely horizontal, the beams reflected by the liquid and the mirror propagate in the same direction, and their fringe patterns coincide as shown in Fig. 7.45a. If the surface is tilted by an angle β, the two reflected beams are mutually deflected by $2\beta d/\theta$, as shown in Fig. 7.45b. The sensitivity that was achieved by the moire deflectometer level is 5.10^{-6} rad.

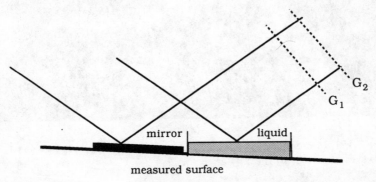

Figure 7.44. The setup of a moire deflectometric level. A collimated beam is simultaneously reflected from a flat mirror and the liquid surface and then passes through two gratings G_1 and G_2. The relative shift of the moire fringes due to a tilt of the surface from horizontal is observed.

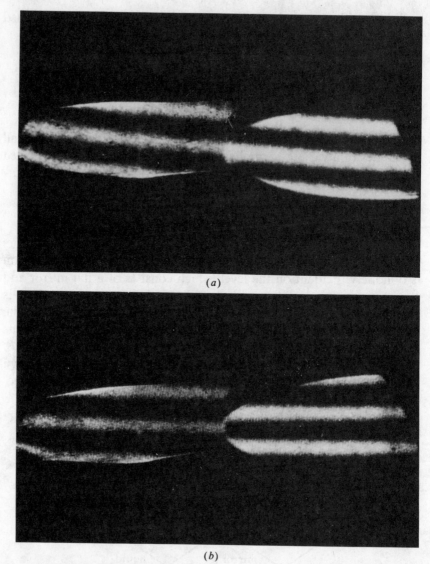

(a)

(b)

Figure 7.45. Moire deflectograms produced by the level of Fig. 7.44 (a) in the levelled position and (b) tilted by an angle.

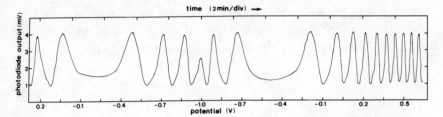

Figure 7.46. A section of a fringe trace of change in electrode length recorded during a linear potential sweep. One period is equal to 0.8μm.

7.6.2. Dimensional Changes of High Surface Area Electrodes [46]

Another application of moire deflectometry in which the measured effect is translated to mirror deflection is measuring the length changes in a graphite sheet electrode immersed in an electrolyte solution subject to electric potential changes. One end of the mirror is held fixed to a hinge while its center is loosely connected to the test electrode. The electrode length changes due to absorption, which is controlled by the voltage applied to the solution and this causes a slight tilt of the mirror. Due to the small changes, one can assume that the deflection angle is proportional to the length change. A collimated beam incident on the mirror is deflected at twice the mirror's tilt angle and the fringe shift can be immediately calibrated in units of electrode length change. Figure 7.46 is a typical chart recording of an elongation–contraction angle which clearly indicates the sweep inversion point at -1.0 V.

7.6.3. Frequency Marker [47]

Moire deflectometry can be used as a frequency marker based on measuring the diffraction angle. For demonstration, a collimated beam of a tunable rhodamine-6G dye laser was diffracted by a 40 lines/mm grating. The first order of diffraction entering the deflectometer, causes a fringe shift of $\lambda d/p\theta$; thus the moire fringe shift will vary with λ. A detector with a limited field of view, due to a narrow slit attached to it, monitors the transmitted intensity vs. the wavelength. In order to calculate the resolving power of the deflectometer, we first calculate the change in the wavelength required to shift the moire fringe one full period p/d. If the pitch of the diffraction grating is p_1, the diffraction angle is λ/p_1 or

$$\frac{\lambda_1 - \lambda_2}{p_1} = \frac{p}{d} \qquad (7.43)$$

or

$$\frac{\lambda}{\delta\lambda} = \frac{\lambda d}{pp_1}. \qquad (7.44)$$

For example, if the gratings gap is 100 cm and $\lambda = 570$ nm, the resolving power is about 6×10^3 using first order diffraction. The upper limit of the resolving power is set by the effective aperture a, i.e., d cannot exceed ap/λ and

$$\frac{\lambda}{\delta\lambda} \leq \frac{a}{p_1} = N, \qquad (7.45)$$

where N is the total number of lines of the grating within the diffracting aperture a (i.e., the resolving power of the grating).

7.6.4. Hybrid Talbot Effect [48]

The same setup as for the frequency marker can be used with a fixed λ to determine the pitch p_1 of a grating. The method is suitable for measuring the pitch of low density gratings (~ 1 line/mm) with high accuracy. The low density grating diffracts the collimated beam into three beams, the zero and the two first diffraction orders. If the distance between the deflectometer gratings d is such that

$$\frac{\lambda d}{p_1} = np \qquad (n \text{ an integer}), \qquad (7.46)$$

a generalized Talbot effect is obtained. The pattern contains optimal contrast moire fringes in addition to the fringes caused by the diffraction, namely,

$$d = npp_1/\lambda. \tag{7.47}$$

An interesting effect occurs when n is an odd half integer. A fringe splitting is obtained, in contrast to normal Talbot effect, where total fringe blur occurs. The hybrid Talbot effect can be used to measure p_1 either by varying d or by rotating the grating (thus tuning p_1) at a fixed grating gap.

REFERENCES

1. J. Krasinski, D. F. Heller, and O. Kafri, Phase Object Microscopy Using Moire Deflectometry, *Appl. Opt.* **24**, 3032–3036 (1985).
2. O. Kafri and I. Glatt, High Sensitivity Reflection–Transmission Moire Deflectometer, *Appl. Opt.* **27**, 351–355 (1988).
3. O. Kafri and A. Livnat, Reflective Surface Analysis Using Moire Deflectometry, *Appl. Opt.* **20**, 3098–3100 (1981).
4. O. Kafri and J. Krasinski, High Sensitivity Moire Deflectometry Using a Telescope, *Appl. Opt.* **24**, 2746–2748 (1985).
5. J. Stricker and J. Politch, Holographic Moire Deflectometry-Method for Stiff Density Fields Analysis, *Appl. Phys. Lett.* **44**, 723–725 (1984).
6. D. B. Rhodes, J. M. Franke, S. B. Jones, and B. L. Leighty, Moire Deflectometry with Deferred Analysis, *Appl. Opt.* **22**, 652–653 (1983).
7. O. Kafri and E. Margalit, Double Exposure Moire Deflectometry for Removing Noise, *Appl. Opt.* **20**, 2344–2345 (1981).
8. O. Kafri, A. Livnat, and E. Keren, Optical Second Differentiation by Shearing Moire Deflectometry, *Appl. Opt.* **22**, 650–652 (1983).
9. O. Kafri and A. Livnat, Second and Third Optical Differentiation by Double Moire Deflectometry, *Appl. Opt.* **22**, 2115–2117 (1983).
10. O. Kafri, Noncoherent Methods for Mapping Phase Objects, *Opt. Lett.* **5**, 555–557 (1980).
11. Z. Karny and O. Kafri, Refractive Index Measurements by Moire Deflectometry, *Appl. Opt.* **21**, 3326–3328 (1982).

12. A. Livnat and O. Kafri, Fringe Addition in Moire Analysis, *Appl. Opt.* **22**, 3013–3015 (1983).

13. I. Glatt, A. Livnat, O. Kafri, and D. F. Heller, Autocollimator Based on Moire Deflectometry, *Appl. Opt.* **23**, 2673–2674 (1984).

14. D. F. Heller, O. Kafri, and J. Krasinski, Direct Birefringence Measurements Using Moire Ray Deflection Technique, *Appl. Opt.* **24**, 3037–3040 (1985).

15. I. Glatt and O. Kafri, Determination of the Focal Length of Non-Paraxial Lenses by Moire Deflectometry, *Appl. Opt.* **26**, 2507–2508 (1987).

16. E. Keren, K. Kreske, and O. Kafri, A Universal Equation for Determining the Focal Length of Lenses and Lens Systems Using Moire Deflectometry, *Appl. Opt.* **27**, 1383–1385 (1988).

17. I. Glatt and A. Livnat, Determination of the Refractive Index of a Lens Using Moire Deflectometry, *Appl. Opt.* **23**, 2241–2243 (1984).

18. O. Kafri and K. Kreske, Flatness Analysis of Hard Disks. *Optical Eng.* **27**, 878–882 (1988).

19. O. Kafri, A. Livnat, and I. Glatt, Temporally Stable Density Patterns, *J. Fluids Eng.* **106**, 257–261 (1984).

20. J. Stricker and O. Kafri, A New Method for Density Gradient Measurements in Compressible Flows, *AIAA Journal* **20**, 820–823 (1982).

21. J. Stricker, E. Keren, and O. Kafri, Axisymmetric Density Field Measurements by Moire Deflectometry, *AIAA Journal* **21**, 1767–1769 (1983).

22. J. Stricker, Analysis of 3-D Phase Objects by Moire Deflectometry, *Appl. Opt.* **23**, 3657–3659 (1984).

23. E. Keren, E. Bar-Ziv, I. Glatt, and O. Kafri, Measurements of Temperature Distribution of Flames by Moire Deflectometry, *Appl. Opt.* **20**, 4263–4267 (1981).

24. E. Bar-Ziv, S. Sgulim, O. Kafri, and E. Keren, Measurement of Temperature Distribution in Methane–Air Flame by Moire Deflectometry, *19th Int. Symp. on Combustion*, The Combustion Inst., 1982, pp. 303–310; Temperature Mapping in Flames by Moire Deflectometry, *Appl. Opt.* **22**, 698–705 (1983).

25. R. C. C. Leite, R. S. Moore, and J. R. Whinnery, *Appl. Phys. Lett.* **5**, 141 (1964).

26. H. L. Fang and R. L. Swofford, The Thermal Lens in Absorbtion Spectroscopy, in *Ultrasensitive Laser Spectroscopy*, D. S. Klinger, Ed., Academic, New York, 1983.

27. J. R. Whinnery, *Acc. Chem. Res.* **7**, 225–231 (1974).

28. I. Glatt, Z. Karny, and O. Kafri, Spatial Analysis of the CO_2 Laser-Induced Thermal Lens in SF_6 by Moire Deflectometry, *Appl. Opt.* **23**, 274–277 (1984).

29. A. Rose, Y.-X. Nie, and R. Gupta, Laser Beam Profile Measurement by Photothermal Deflection Technique, *Appl. Opt.* **25**, 1738–1741 (1986).

30. I. Glatt, R. J. Kerl, and C. K. N. Patel, unpublished results.

31. C. K. N. Patel, E. T. Nelson, and R. J. Kerl, Vibrational Overtone Absorption in Solid Hydrogen, *Phys. Rev. Lett.* **22**, 1631–1635 (1981).

32. I. Glatt, R. J. Kerl, and C. K. N. Patel, Observation of Fourth Vibrational Overtone of Hydrogen, *Phys. Rev. Lett.* **57**, 1437–1439 (1986).

33. C. K. N. Patel, E. T. Nelson, and R. J. Kerl, *Phys. Rev. Lett.* **22**, 1631 (1981).

34. L. Horowitz, Y. B. Band, O. Kafri, and D. F. Heller, Thermal Lensing Analysis of Alexandrite Laser Rods by Moire Deflectometry, *Appl. Opt.* **23**, 2229–2231 (1984).

35. T. Chin, R. C. Morris, O. Kafri, M. Long, and D. F. Heller, Athermal Nd : BEL, Conference on Lasers and Electro-Optics, 9–13 June, 1986, San Francisco, Calif., paper no. WM2, 212-14.

36. E. Keren, A. Livnat, and O. Kafri, A Generalized Theory for the Optical Transfer Function—Ray Optical Approach, *J. Opt. Soc. Am. A* **5**, 1213–1226 (1988).

37. I. Glatt and O. Kafri, Fringe Contrast Measurements in Moire Deflectometry Using Vibrating Gratings, *Appl. Opt.* **24**, 2468–2470 (1985).

38. *Melles Griot Optics Guide* **3**, 130–131.

39. Z. Karny, S. Lavi, and O. Kafri, Direct Determination of the Number of Transverse Modes of a Light Beam, *Opt. Lett.* **8**, 409–411 (1983).

40. I. Glatt, A. Livnat, and O. Kafri, Beam-Quality Analysis of Surface Finish by Moire Deflectometry, *Exp. Mech.* **24**, 248–251 (1984).

41. O. Kafri, H. Samelson, T. Chin, and D. F. Heller, Moire Modulation Transfer Function of Alexandrite Rods and Their Thresholds as Laser, *Opt. Lett.* **11**, 201–203 (1986).

42. I. Glatt and O. Kafri, Analysis of the Turbulent Mixing of Liquids by Moire Deflectometry, *Chem. Eng. Sci.* **39**, 1637–1638 (1984).

43. D. L. Fried, Limiting Resolution Looking Down Through the Atmosphere, *J. Opt. Soc. Am.* **56**, 1380–1390 (1966).

44. E. Keren, I. Glatt, and O. Kafri, Propagator for the Modulation Transfer Function of a Wide-Angle Scatterer, *Opt. Lett.* **11**, 554–556 (1986).

45. A. Livnat and O. Kafri, Level Based on Moire Deflectometry, *Rev. Sci. Instrum.* **53**, 1779–1781 (1982).

46. Y. Oren, I. Glatt, A. Livnat, O. Kafri, and A. Soffer, The Electrical Double Layer and Associated Dimensional Changes of High Surface Area Electrodes as Detected by Moire Deflectometry, *J. Electroanal. Chem. and Interfacial Electrochem.* **187**, 59–71 (1985).

47. O. Kafri, D. Meyerstein, and Z. Karny, Frequency Marker Based on Moire Deflectometry, *Opt. and Lasers in Eng.* **4**, 55–61 (1983).

48. O. Kafri, E. Keren, and A. Livnat, Measurement of Low Density Transmission Gratings by Moire Deflectometry: Hybrid Talbot Effect, *Appl. Opt.* **26**, 28–29 (1987).

Appendix:

Aberration Analysis Using Moire Deflectometry

A.1. BACKGROUND

Moire deflectometry was also found to be useful for the analysis of lens aberrations [1]. By aberrations we denote the departure of rays from the course predicted by the paraxial theory. In spherical lenses they are caused by the rays passing through the lens close to the edge (marginal rays), and they appear as a distortion or blur of the image produced by the lens. Some aberrations result from imperfections within the lens, some result from the fact that the lens is spherical rather than parabolical, whereas others are caused by a misalignment of the lens in the optical system.

We show how to identify and interpret the most common aberrations that exist to some degree in all single element lenses. We discuss only the third order aberrations, so called because they provide an improved third order approximation to geometrical optics [2, 3] by expressing $\sin \theta$ as $\theta - \theta^3/3!$ (instead of the approximation $\sin \theta \sim \theta$ used in the paraxial theory). The main third order aberrations are:

Spherical aberrations: These aberrations correspond to a dependence of the focal length on the lens aperture. For a converging lens, marginal rays will be focused closer to the lens than the paraxial rays, and instead of a sharp focus, a caustic will be formed.

Comatic aberrations (coma): Coma occurs because the principal planes of a lens are not planar, as assumed in the paraxial approximation, but curved. Therefore, when the lens is tilted, the effective focal length of the off-axis rays differs from that of the rays lying on-axis. Hence, when the ray's incidence is oblique, a comet-like flare is observed at the focus.

Astigmatism: This aberration occurs when the focal length of the lens in the meridional plane, i.e., the plane containing the chief ray (the ray passing through the lens center) and the optical axis, differs from the focal length of the lens in the sagittal plane, the plane perpendicular to it. Astigmatism occurs when the lens is tilted with respect to the incident beam. The meridional rays, which are tilted more than the sagittal rays, will have a shorter focal length and the astigmatic difference will increases as the rays incidence becomes more oblique. If the lens is tilted with respect to the optical axis, astigmatism occurs when the focus in the direction of tilt is at a different location than the focus in the direction perpendicular to the tilt.

A.2. EXPERIMENTAL TECHNIQUES

As derived by Kingslake [4] in 1925, the wave front distortion due to primary aberrations is given by the general expression

$$W(x, y) = A(x^2 + y^2)^2 + By(x^2 + y^2) + C(x^2 + 3y^2)$$
$$+ D(x^2 + y^2) + Ey + Fx, \qquad (A.1)$$

where each term represents a different type of aberration as follows:

A = spherical aberrations D = defocusing
B = coma E = tilt about the x axis
C = astigmatism F = tilt about the y axis

The first three, spherical aberrations, coma, and astigmatism, are actual aberrations within the lens. Tilt and defocusing result from an error in the lens positioning within the system.

Moire deflectometry studies aberrations through their effect on the lateral derivatives of the wave front, either $\partial W(x, y)/\partial y$ or $\partial W(x, y)/\partial x$. It is assumed that a wave front resulting from a nonaberrating lens is perfectly planar, i.e., the lateral derivatives are 0. It must be noted that since derivatives are the measured quantity, deflectometry is not sensitive to the tilt terms, which add only a constant phase to the deflectogram.

The configuration used for testing aberrations with moire deflectometry is shown in Fig. A.1a. The lens is placed in the path of the collimated beam. It focuses the beam on the mirror and this point source will then be recollimated by the lens on the return trip. For

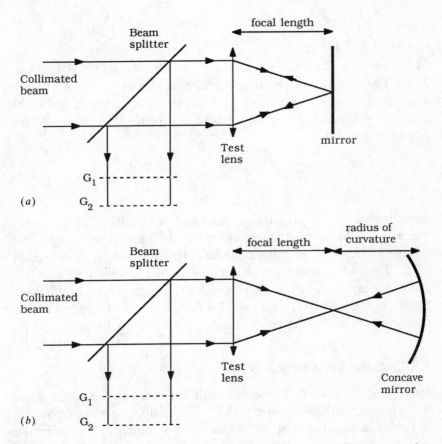

Figure A.1. (a) Setup for measuring lens aberrations. (b) Altered setup for measuring coma.

measuring coma, the slightly altered setup of Fig. A.1b, must be used. This is because coma, an asymmetric aberration, is cancelled in the first setup.

A.2.1. Defocusing

Defocusing is the simplest type of aberration and it occurs when the lens is not positioned exactly at the focal distance f, but shifted slightly by δz. Therefore, the resulting beam is no longer collimated, but has the parabolic wave front

$$W(x, y) = D(x^2 + y^2), \quad \text{where } D = \frac{\delta z}{2f^2}. \qquad (A.2)$$

The deflection angle ϕ_x measured by the deflectometer is $\partial W/\partial x = 2Dx$. Therefore, in the infinite fringe mode we observe straight equidistant fringes of period p' (see Sec. 7.3.2). When p is the gratings' pitch and d is distance between them, p' is related to the defocusing D by

$$D = \frac{\phi}{2p'}, \quad \text{where} \quad \phi = \frac{p}{d}. \qquad (A.3)$$

Rearranging the terms in Eq. (A.3) and substituting a/N for p', where a is the aperture size and N is the number of fringes, we receive $f^2/\delta z = ad/Np$, the same focal length equation as in Section 7.3.2. This is because a small amount of defocusing produces the same lensing effect as a weak lens of focal length $f^2/\delta z$. In the finite fringe mode, tilted fringes are formed and the direction of tilt determines whether the lens is inside or outside the focus.

A.2.2. Spherical Aberrations

Since most lenses are fabricated with spherical surfaces, the wave front is not parabolic as in Eq. (A.2). Expanding the wave front into a Taylor series, the first correction in a symmetric lens is in the fourth order:

$$W(x, y) = Ar^4 + Dr^2, \quad r^2 = x^2 + y^2. \qquad (A.4)$$

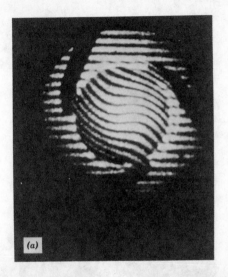

 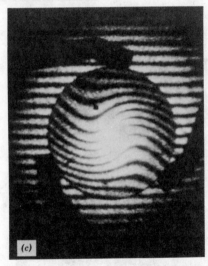

Figure A.2.   Finite fringe deflectograms of spherical aberrations. (a) Paraxial focus. (b) Marginal focus. (c) Best focus.

Figure A.3. Infinite fringe deflectograms of spherical aberrations. (*a*) Paraxial focus. (*b*) Marginal focus. (*c*) Best focus.

A is the spherical aberration coefficient and the deflection angle in the x direction is

$$\phi(x, y) = 4Ax(x^2 + y^2) + 2Dx. \tag{A.5}$$

The focus is attained when $\phi(x, y) = 0$, i.e., when the beam is recollimated after passing through the lens twice. For the paraxial rays passing through the lens near its center ($r \approx 0$), the focus occurs at $D = 0$ (Figs. A.2a and A.3a). Note that in the infinite fringe pattern the fringes are not continuous with the reference fringes.

For the marginal rays passing through the lens near its edge ($r \approx a/2$), the focus occurs at $D = -2A(a/2)^2$. Figure A.2b shows a lens at the marginal focus. Note that the fringes show zero deflection at the center ($r \sim 0$) and the margins ($r \sim a/2$) of the lens. These fringes are continuous at the lens surrounding interface. In the infinite fringe mode, a series of closed circular fringes will be formed (Fig. A.3b). The best focus is found one-quarter of the distance from the marginal focus to the paraxial focus at $D = -A(a/2)^2/2$. This situation is visualized as a unification of the two images of the aperture and can be seen in Fig. A.2c and Fig. A.3c.

The coefficient A can be derived from D, where δz is the distance between the marginal and paraxial foci. An alternative way uses the infinite fringe pattern obtained at the paraxial focus (Fig. A.3a). Given $\phi = 4Ax^3$ and $\phi_{incr} = p/d$, A can be derived from the slope of $(x^3)_m$ vs. m where m is the fringe index and $\phi_m = m\phi_{incr}$.

A.2.3. Coma

Differentiating Eq. (A.1), we obtain the two ray deflection equations

$$\frac{\partial W(x, y)}{\partial y} = \phi_y(x, y)$$

$$= 4Ay(x^2 + y^2) + B(x^2 + 3y^2) + (6C + 2D)y + E,$$

$$\frac{\partial W(x, y)}{\partial x} = \phi_x(x, y)$$

$$= 4Ax(x^2 + y^2) + 2Bxy + (2C + 2D)x + F. \tag{A.6}$$

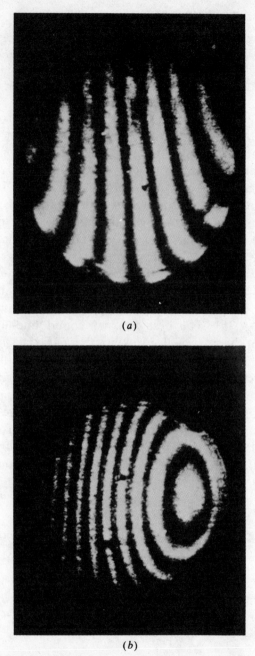

Figure A.4. Infinite fringe deflectograms of coma. (*a*) Ellipsoidal fringes. (*b*) Hyperbolic fringes.

The two second terms, $B(x^2 + 3y^2)$ and $2Bxy$, represent coma. The infinite fringe deflectograms from comatic aberrations appear completely different in the direction of tilt and the direction perpendicular to the tilt. The coma term in the direction of tilt, i.e., $B(x^2 + 3y^2)$, appears as ellipsoidal fringes with the axes ratio $\sqrt{3}$ to 1 (Fig. A.4a). In the direction perpendicular to the tilt, the coma fringes are hyperbolas (Fig. A.4b).

A.2.4. Astigmatism

Notice that the defocusing and astigmatism terms are similar. This is expected since astigmatism refers to an asymmetric defocusing due to the tilt of the lens. If the lens is tilted with respect to the x axis, the deflections are represented by the two third terms in Eq. (A.6), $(6C + 2D)y$ and $(2C + 2D)x$. If the lens is tilted with respect to the y axis, the equations are interchanged. As with symmetrical defocusing, in the infinite fringe mode straight equidistant fringes are observed in both directions. The astigmatism results in a difference in their separation p' of

$$p_\perp = \frac{\phi_{\text{incr}}}{6C + 2D} = \frac{p}{2d(3C + D)}$$

and

$$p_\parallel = \frac{\phi_{\text{incr}}}{2C + 2D} = \frac{p}{2d(C + D)}. \tag{A.7}$$

Astigmatism can also be detected in the finite fringe mode and the tilt of the fringes relative to the unperturbed fringes will change as the deflectometer is rotated. At the sagittal focus, one infinite fringe is observed when measuring the gradients in the direction perpendicular to the tilt, i.e., $p_\perp = \infty$ and $D = -3C$. At the meridional focus, infinite fringe is observed when measuring the gradients in the direction of the tilt of the lens, i.e., $p_\parallel = \infty$, or rather $D = -C$. To find the medium focus we set $(6C + 2D) = -(2C + 2D)$, and therefore $D = -2C$. Here we will find that $p_\perp = -p_\parallel$ and the same number of fringes is observed in both directions. Switching to the finite mode we find that the fringes are rotated in opposite directions. The astigmatism coefficient C for a given tilt angle is

$$C = \frac{1}{4}\left(\frac{1}{p_\perp} - \frac{1}{p_\parallel}\right)\frac{p}{d}.$$

(A.8)

A.3. NORMALIZATION OF THE COEFFICIENTS

The values of A, B, C, and D that are found in texts are often normalized so that they are easier to interpret. Using the normalization technique of *Optical Shop Testing* [2], the aperture radius is normalized to be unity and the coefficients A', B', C', and D' represent the number of wavelengths of aberrations. The normalization factors are $A' = A(a/2)^4/\lambda$, $B' = B(a/2)^3/\lambda$, $C' = C(a/2)^2/\lambda$, and $D' = D(a/2)^2/\lambda$, where a is the lens diameter and λ is the wavelength.

REFERENCES

1. K. Kreske, E. Keren, and O. Kafri, *Experimental Handbook for the OMS-400*, Rotlex Optics Publication, 1988.
2. D. Malacara, *Optical Shop Testing*, Wiley, New York, 1978, Secs. 2.3, 4.3, 9.2, and A.3.
3. E. Hecht and A. Zajac, *Optics*, Addison-Wesley, Reading, MA, 1974, Sec. 6.3.
4. R. Kingslake, The Interferometer Patterns Due to the Primary Aberrations, *Trans. Opt. Soc.* **27**, 94 (1925–1926).

Author Index

Numbers in *italic* indicate pages on which the complete references are listed.

185

Subject Index

RETURN   PHYSICS LIBRARY
TO➡️   351 LeConte Hall
LOAN PERIOD